Guarantee the Sex of YOUR BABY

Guarantee the Sex of YOUR BABY

CHOOSE A GIRL OR BOY
USING TODAY'S 99.99% ACCURATE SEX SELECTION TECHNIQUES

ROBIN ELISE WEISS, LCCE

Foreword by Jeffrey Steinberg, M.D.

Ulysses Press

Published by: Ulysses Press
P.O. Box 3440
Berkeley, CA 94703
www.ulyssespress.com

ISBN10: 1-56975-570-1
ISBN13: 978-1-56975-570-9
Library of Congress Control Number: 2006907920

Printed in Canada by Transcontinental Printing

10 9 8 7 6 5 4 3 2 1

Acquisitions Editor: Nicholas Denton-Brown
Managing Editor: Claire Chun
Editor: Mark Woodworth
Editorial and production staff: Lily Chou, Lisa Kester, Matt Orendorff, Elyce Petker
Index: Sayre Van Young
Cover design: what!design @ whatweb.com
Cover photo: © iStockphoto.com/Amanda Rohde

Distributed by Publishers Group West

With special love and gratitude to my husband, Kevin,
and our sons and daughters:
Hilary, Benjamin, Isaac, Lilah, Owen, Clara, and Ada.

Table of Contents

Foreword

For the first time in history, couples contemplating pregnancy have the option of pre-selecting, with virtually 100 percent certainty, the sex of a child resulting from a pregnancy they may be planning. For those who may be considering sex selection of any type, Robin Elise Weiss's *Guarantee the Sex of Your Baby* offers an extremely well-researched, accurate, complete and long-overdue new guide to the complex modern science and techniques involved in modern-day sex selection. Having now provided sex selection services to more couples than any other program worldwide, our center was excited to hear about Robin's new work and have found it to be the most valuable resource for couples who are beginning to learn about and participate in modern sex selection. As Robin details in the book, couples today may choose from a variety of sex selection options that ranges from the most modern preimplantation genetic diagnosis (PGD) treatments, offering 100 percent accuracy, to long-standing, widely practiced but totally ineffective methods that offer no more chance of obtaining a baby of a desired sex than nature's own 50-50 (or even lower) odds. *Guarantee the Sex of Your Baby* offers an important overview of the history of the scientific advances that have occurred in sex selection. It aids those considering the process in becoming fully educated about what is in store for them. And this education is vitally important. It provides a safety net that will help couples ask the right questions and make the right decisions.

The book provides a level of accurate counseling and guidance that those of us performing these modern procedures wish all potential sex selection parents were able to take advantage of. Every week, our center sees dozens of patients who are chasing their dream of a baby of a desired sex and attempting to balance their families. I personally spend the initial part of a new-patient interview dispelling the myths that abound concerning achieving a desired sex. Most of these patients have spent days, and often months, researching available technologies. But the available information has been confusing and, in many instances, misleading and often wrong. *Guarantee the Sex of Your Baby*, with its easy-to-read, easy-to-understand format, will abate the state of overwhelmed confusion and encourage full understanding of what science can now offer.

I recall my initial contact with sex selection techniques as a resident in obstetrics and gynecology at Rush Medical College and Presbyterian–St. Luke's Hospital in Chicago. I had been assisting a young woman and her husband through a lengthy labor and delivery. The husband was calling the babysitter at home every 30 minutes to check up on the couple's other children. Relatively new to the field, I felt good assisting in the birth of a baby whose waiting family seemed to be a very loving and welcoming one. All went well, and a healthy baby boy was born. As I did rounds the next morning to check up on the mothers who had deliveries the night before, I was surprised to find this same young mother weeping quietly in her hospital bed. I asked what she was feeling and learned that the couple had three healthy children at home, all boys. My patient confided that she and her husband had bought the most popular book available allegedly explaining how to achieve a baby of a desired sex. They practiced the rigors and acrobatics of the "formula" religiously. She was certain that they would deliver a baby girl whom they so longed for. I wondered why she was so surprised to see the baby boy. She had undergone ultrasound studies, one of them at a point in the pregnancy where the sex should have been obvious. I inquired about this. She told me that she had asked not to be told the sex with her ultrasound studies. She "wanted to feel what it was like to carry a girl...even if it turned out not to be true." It was then that I decided to investigate the "miracle" book she had relied upon. Fresh out of medical school, and with science deeply engrained in my training, I was astonished at what I learned. The "method" my patient had relied upon and placed her faith in had not one shred of scientific credibility! As I delved further, I learned that I should not have been surprised. In those days, there was *no* method that offered any sort of scientifically credible alteration of the chances of achieving a desired sex. Despite this, millions of copies of various "method" books, instruction manuals, and formulas were being sold.

Now, years later, there has been a renaissance in the science of sex selection. One-hundred-percent success is readily achievable and we offer it on a daily basis. While sex selection has been legislated out of use in most of the world, the United States, in its rigorous defense of reproductive freedom and choice, continues to not abridge a couple's right to make their own reproductive choices. Yet, in spite of the huge advances in sex selection success, I still remain frustrated by the number of patients who are quietly disappointed by the same outdated techniques I found years ago to be essentially worthless.

Whether you are content to elect one of the modern techniques, with success ranging from 70 to 80 percent, all the way up to the scientifically validated and proven PGD process at 100 percent, or prefer to rely on pure luck with a copy of one of the outdated method books, you could not make a better choice than reading *Guarantee the Sex of Your Baby*. It is destined to become the new gold standard in sex selection education for those truly interested in finally balancing their families with the arrival of that elusive baby girl or boy.

Jeffrey Steinberg, M.D., FACOG
Medical Director, The Gender Selection Program
The Fertility Institutes; Los Angeles, Las Vegas, New York, Guadalajara

Introduction

"Girl or boy?" Whether you have just given birth to your first or fifth baby, that is the question everyone wants to hear you answer. If you're having an ultrasound done, the first thing your friends and family ask is *not* "Is the baby healthy?" but rather, "Is it a boy or a girl?" Today's society has become inordinately obsessed with the sex of the baby. Our desire to know even if the child is healthy is surpassed by this assignment of sex.

The preference to select the sex of your child predates the ability to accurately do so by a very long time. Since the beginning of time, men and women have done many things to influence the sex of their offspring. The desire is not new, nor is it likely to wane anytime in the foreseeable future. Although each generation has its own spin on sex selection, the generation of young parents having babies today will go down in history as the one for whom it became both a possibility and a reality for nearly all of them to accurately choose their babies' sex.

There are new technologies that allow couples to choose with astonishing accuracy the baby of their dreams. For today we have the revolutionary ability to sort the male from the female sperm to increase the odds of having a baby of a desired particular sex. You can actually pick and choose a specific embryo based on a series of criteria, including the sex of the embryo and the health of that embryo. These amazing advances were only dreamed of in the not-so-distant past.

Yet this is a far cry from what some mockingly call "designer" babies. There is no choice of eye color, hair color, or hand preference. You don't simply fill out a form and swipe your credit card, then pick up your prewrapped "perfect" baby at the drive-through window on your way out. There's much more to the process than that.

Some families choose their baby's sex; having children of both sexes within the family is what is defined as family balancing. The practice may call for adding a girl to a long line of sons, or selecting to have one child of each sex. There are also those who may choose to add a certain sex to try to even the numbers—for example, if they already had at least one child of each sex, but ended up with more of one than the other. Families choosing sex selection for family balancing purposes make up the majority of those who are investigating sex selection.

For other families, there may be this deep desire for a certain-sex child so that it won't be prone to a sex-linked genetic disorder that they're aware of. Having one of these disorders, like Duchenne muscular dystrophy (DMD), Tay-Sachs disease, sickle cell disease, thalassemia, Huntington's disease, or hemophilia, would mean that a child of a certain sex would suffer a lifelong course of medical treatments, pain, and maybe even premature death. When such parents are empowered to select the sex of their baby, they can prevent this pain and suffering in their off-spring—a gift that most reasonable mothers and fathers would want to be able to bestow on a newborn.

These preferences have, of course, been debated for years as an ethical concern. The side speaking against sex selection thinks that nature has a plan. They believe that the sex of children is predetermined and that diversion from that predetermined path can seriously harm the balance of the earth and the cultures living on it.

On the other side of the argument are proponents of sex selection who say that by using the high-tech, effective methods of sex selection we as a society can reduce the numbers not only of ill children who are destined to suffer a life of affliction, but also of unwanted children who may not have a pleasant life in foster families, or worse. Others on this side are quick to point out that sex selection can also reduce abortions for sex selection purposes, and even infanticide in areas where it's still practiced.

The debate over sex selection, however, is not the subject of this book. The Resources section provides many references and websites where you can learn more on this topic; I encourage you to read these thoughtful documents from both sides of the issue so that you can make informed decisions about what would best meet your family's needs.

Regardless of the reasons you might have for investigating sex selection, it remains a personal issue, no matter what justification you use for wanting to do it. The good news is that many opportunities are available these days for sex selection that are both safe and amazingly accurate. No longer are you required to take your best guess and throw caution to the winds while using anecdotal evidence, or perhaps odd or even unsafe methods, to try to have a baby of the sex that you desire.

One of the hottest issues within the sex selection realm is the arena of privacy. The doctors, lab directors, nurses, and others involved with the facilities that do sex selection are all quite used to the idea of maintaining your privacy and are entirely respectful of your desires and wishes. Both before and after birth, there's no reason to let others know about the facts surrounding your baby's conception, unless you and your spouse or partner choose to tell.

So while the desire to be able to choose the sex of your children is not original, the great news is that there is new technology available that can virtually assure you that you'll have the baby of the sex you prefer. And this technology works, regardless of the reason behind the procedure being done.

This book is designed as a tool and resource guide to help you look at *all* your choices in sex selection. While it will include a discussion of most major methods of sex selection, its goal is to help you "get it right" nearly every time by using the best and latest methods of sex selection. By "best" I mean the ones that are the most accurate in providing you with that baby girl or boy of your dreams.

Robin Elise Weiss, BA, CLC,
ICCE-CPE, CD(DONA), LCCE, FACCE

PART ONE

The History of Sex Selection, Family Planning, and Reproduction

1

A Brief History of Sex Selection

The yearning to influence the sex of a baby is an ancient one. The historical accounts of sex selection show examples that, in light of today's science, seem barbaric and strange, yet were exactly what the doctor of the day ordered. As you read some of these examples, imagine asking your husband or wife to try some of these ancient sex selection rituals. Suddenly, merely asking for a semen sample seems quite polite and civil....

Some "physicians" or shamans or village elders in the distant past would recommend things for people to ingest in order to have a girl or a boy. These concoctions might be a mixture of nasty-tasting herbs or even a selection of not truly edible items, depending on the working theory of the prescriber.

For years, theories abounded on how to produce a baby boy that border on what, today, we would call ridiculous. How could bringing an ax to bed actually ensure that a male child would be conceived? I suppose that notion went along with the man wearing his boots to bed—a "manly" act produces manly issue.

In ancient China, it was believed that the mother was in control of the baby's sex. In fact, the woman had so much control that she could supposedly change the sex of the baby she was carrying during pregnancy. She was purportedly able to do this, at will, up until the third month of her gestation.

Theories also surrounded the moon and the lunar calendar. The moon has often been thought of as a fertility symbol. In ancient times, having sex during the full moon would help ensure that your baby would be a boy. Other theories held during the same era would say that this was wrong, that sex during a full moon would only produce a girl. If you wanted a baby boy, you needed to wait until the quarter moon.

Herein lies part of the problem with ancient sex selection methods and theories—they rarely worked as they were supposed to, meaning that sometimes you got "lucky" and sometimes you didn't. When you didn't get lucky, you still had a baby, even if it was of the "wrong" sex.

The contradictions abounded from method to method. Try it this way for a girl, one theory says, but then a parent would hear about another theory that says precisely the opposite. More often than not, of course, these methods were not based on scientific fact, nor did they reliably produce the results that were desired.

The truth is that even today, these methods and theories continue to be used, and they can actually do great mental and sometimes even physical harm. The harm comes when parents' deep-seated desire or dream is shattered after failure of the "perfect" attempt at a low-tech sex selection method. Harm can also be caused by ingesting some of the supplements advised in folk methods, or by altering the natural flora of the vaginal tract with a variety of strange douching concoctions that are sometimes recommended. One mother recalls:

> "I read that I was supposed to douche with lime juice. Then I read about someone trying to figure out how to get the pulp out of the juice. That just can't be healthy for my body."

But physical harm in sex selection is nothing new. In ages past some subscribed to the theory that since a male had two testicles, the key to sex selection lay in that area. While not far from the truth of how the sex of a baby was determined, they missed the mark in their cures. Some ancient practitioners would actually have men tie something tight around one testicle or the other, depending on which sex baby was desired. If you wanted a boy, you would tie up your left testicle, theoretically allowing only sperm from the right testicle to be ejaculated.

These ancient attempts were usually ridiculous and even potentially harmful to the participants. Worse than that, these ancient theories and practices often failed. When they failed the results were often even more disastrous, particularly for the baby in question, as it often meant infanticide.

Thankfully, today the killing of babies is not a common practice when a baby is born and is of the opposite sex than what parents had planned or hoped. And now technology is available that virtually guarantees parents the sex of the child for which they are planning and hoping. It is these new technologies that we will focus on in this book, but before you can understand where we are going with new technologies in sex selection, you need to know where we have been.

The Politics of Sex Selection

Boy or girl? What's the big deal—all we really want is a baby, right? While most couples considering sex selection would dismiss the thought that having a healthy child is less important than having a child of a chosen sex, many parents still think that sex selection is a really big deal. It may be considered a taboo subject to even talk about.

When something becomes difficult to even discuss within a family, it can be challenging to sort out the potential conflicts and feelings surrounding the issue. Many families have multiple issues to deal with when it comes to the subject of sex selection as it applies to their families.

One of the more commonly discussed reasons for sex selection given is to avoid medical problems for the baby. These genetic issues usually surround an X-linked disorder inherited by the baby. Yet by using today's technology to select a specific sex, the disorder can be avoided nearly 100 percent of the time. We also have the technology to screen an embryo before it is even implanted in the uterus, not only against multiple genetic disorders but for the sole purpose of sex selection as well.

While selecting by sex is a highly emotional issue for most, there is a financial dimension to it as well, both before and after birth. The financial costs of raising a child with a severe disability are quite extensive. This strain on the family can cause many problems, including feelings of guilt should the parents not be able to afford the best care they would desire for their child. The emotional costs are too numerous to even comprehend, let alone describe here. As one mother considering sex selection for genetic reasons says:

> "I can't even begin to calculate the financial drain on our family that having an ill child would be, nor can I fathom the emotional drain of choosing to have no children, or having a child with a debilitating and fatal illness."

Even the most stringent critics of sex selection avow that the practice is acceptable for genetic reasons. The real surprise? Genetic issues make up only a reported 10 percent of all sex selection requests, according to some practitioners.

That leaves about 9 out of 10 sex selection cases that are done for other reasons. These other reasons vary, but generally can be grouped into a category called family balancing. Typically, we think of family balancing as meaning achieving a family with children who are all the same sex, whether several sons or several daughters. The real surprise is that there is no one typical profile of parents who select their next baby by sex alone.

While some families have many children all of one sex, others have only one child. In the end, the numbers of girls and boys will be nearly the same, even if there are large differences among specific families—across the board, even if specific families have varied assortments of boys or girls. The parents with only one child may want merely two children and yet they desire to have the joy and pleasure, the experience, of raising both a son and a daughter. "They are buying an experience," says one physician.

Women will tell you stories of their dreams, beginning when they were little girls fantasizing about their future children. These women wistfully talk about how they had such big plans. They speak of shopping trips with their daughters, wedding dances with sons; of painting rooms pink or blue, using names they have had chosen years earlier with their husbands, and of picking out clothes and other things for their dream babies.

The pain that these mothers feel when they fail to bear a child of the "right" sex is more than just emotional angst, as strong as that might be. The longing and desire that they hold in their hearts can translate into real physical pain. One mother even spoke about not having the daughter of her dreams in this way:

> "Where is she? Where is the daughter I always knew I'd have? I've put a lot of thought into this and if she's not here, then she's lost. Gone. It's like a mental miscarriage, of sorts...something that I need to grieve."

Lest you believe that sex selection is always the dream of the mother, listen to some of the fathers. One dad smiles as he recalls:

> "I was always friends with women growing up. I just knew I'd have daughters. I couldn't wait to have that relationship. I'm thrilled to be the father of a baby boy, but it's just plain weird to me. I've had a lot more learning to do to be the parent of a baby boy."

And when those dreams look like they are slipping away from the expectant and hopeful parents, all because their children aren't coming out as they had mentally planned, some decide to take fate into their own hands. The good news is that today, they are able to actually do something productive when it comes to sex selection.

Preference for a Specific Sex

The thought that boys are more popular than girls is not necessarily false. We have all heard about the heirs "required" by monarchies over the centuries and right up till today. The truth is that in nature more baby boys are born than girls. In fact, about 1,050 baby boys are born as against every 1,000 baby girls.

Historians talk about the need for boys in certain communities or tribes. A boy was more useful and welcome in a group that depended for its survival on vigorous hunting or strenuous farming. A young man was felt to be strapping and strong, whereas a female was seen as more of a liability and less productive. Certainly this has been told to us often enough to believe it.

However, the days of boys being more popular than girls may be over. In a poll conducted on *pregnancy.about.com*, 91 percent of the more than 1,000 respondents said that if they had a sex preference at all, it was for a baby girl. Dr. Ronald Ericsson, a sex selection pioneer, hypothesizes:

> "Women are primarily the ones that decide whether or not to have another child. Those of age to reproduce have a much different opinion of themselves and their position in society than did their mothers and grandmothers. Therefore, they are motivated to have daughters, as they see a bright future for them."

When asked about how long this current trend will last, the doctor offers this: "The era of wanting a first-born male is gone, not to return." While only time will tell for certain, the desire of parents to have a baby girl remains extremely popular today. Will this swing the balance out of proportion, as it supposedly has in favor of males in other countries?

2

A Crash Course in Reproduction

Reproduction is a basic bodily function for most people. Unless couples are trying to get pregnant or are having fertility issues, getting pregnant is something the majority of families put little thought or effort into. They want a baby? They have a baby. Issues of substance that concern them may be of a genetic variety, such as the question of gene disorders, or issues of family balancing to even out the numbers of males and females in one family, or simply issues of preference for one sex over another.

Before you can go on to understand how sex selection works, you must first understand the basics of the human reproductive cycle. With a working knowledge of reproduction, you can positively influence your ability to get pregnant. You also need to understand the basics of the menstrual cycle, including what ovulation is and when it's likely to occur. You will also want to know how the male contribution, sperm, to the pregnancy affects your chances of conception, as well as how the female contribution, an egg, does as well. This information will help you understand how each method of sex selection supposedly works—or doesn't work, as in the case of many of the low-tech methods used to select for a specific sex.

The Human Sperm

Sperm is at the heart of sex selection. You may have heard of "girl sperm" and "boy sperm." This is how the average person thinks of the genetics behind the differences in sperm. Each sperm contains the genetic makeup, when combined with an egg, to become a boy or a girl. It's the sperm that will decide which the fertilized egg will be, because of the sperm's varying contribution.

Sperm is made in the testes of a man in a process known as spermatogenesis. Sperm are continuously produced from puberty until a man dies. Theoretically, every man can impregnate a fertile, egg-bearing woman from his puberty until he dies. Although it was once believed that there were no ramifications of aging on the sperm or reproductive health, today we know that aging in fact causes a decrease in a man's fertility and poses increased risks of miscarriage, whatever the age of the woman.

The cycle to produce a mature sperm (spermatozoa) takes about three months. This means that having a drink today will affect your sperm quality and quantity for some 90 days. The same is true for medications and illness. This is why what a man does in the preconception period, and just before it, does affect how likely the woman is to conceive, based on the quality and quantity of his sperm.

The testes are located outside the man's body, in the scrotum, the fleshy sack underneath the penis that cradles the testicles. The scrotum allows the testes to regulate the temperature of the sperm by raising and lowering themselves to keep them slightly below a normal body temperature. So when the testes are overheated, they're lowered away from the man's body. When they're too cool, the testes are pulled closer for warmth. This keeps them at a constant, near-perfect temperature.

You may have heard that if a man wants to be fertile he should wear boxers. This is true, because tight underwear can keep the testes too close to the body and thus prevent the scrotum from being able to regulate their temperature. And this can decrease a man's fertility.

Ejaculation is the process by which sperm, mixed with other fluids (known as semen), is released at orgasm through the opening at the end of the penis. Sperm that isn't released through ejaculation is reabsorbed back into the body.

A sperm is divided into two parts: a head and a tail. The tail is what helps propel the sperm through the cervical secretions toward the egg in the race to conceive. The sperm contains precisely half the genetic makeup of a baby. The chromosomes are stored in the head of the sperm. It contains 22 chromosomes plus an X (girl) or Y (boy) chromosome. It is this X or Y chromosome that will determine the sex of the baby from the moment of conception.

The Human Egg

The human egg is absolutely necessary for human reproduction. Without the human egg, there can be no conception—and no human reproduction. The egg is required for conception, no matter how you get pregnant.

The egg, or ovum, is produced in a process called oogenesis in a woman's ovary. A baby girl is born carrying all the eggs she will ever have in her life. Once a female reaches puberty, her eggs begin to mature through a hormonal process. As she begins to have menstrual cycles, typically about once each month, usually one egg is prepared to be released each cycle in a process called ovulation.

Each egg released from the ovary contains exactly half the chromosomes needed to create a baby, matched by the number of chromosomes contained in the sperm. The egg is only able to provide a new life with an X chromosome. The meaning of this is that the female alone *cannot* determine the sex of a child, since the eggs will only be X bearing. This certainly breaks the myth that women are in control of the sex of their baby! In ancient times, painful as it is to think about, women were put through torture and sometimes murdered for failure to produce a male heir. Thankfully, this practice is also not a common one today.

The Menstrual Cycle

The process known as the menstrual cycle builds the lining of the uterus to accept a fertilized egg, produces hormones that allow the process of ovulation, and finally sheds the uterine lining if a pregnancy fails to occur. The menstrual cycle is often talked about as being a 28-day cycle. While this is certainly true for many women, it is by no means the cycle length for every woman.

The first day of the menstrual cycle is always calculated from the first day of bleeding, also known as a woman's period. The number of days in a menstrual cycle can vary from woman to woman and cycle to cycle. The number of days in your cycle is a critical number to help maximize your chances of conception. It is even more vital to a family using sex selection, for no matter what technique they use, they must be able to pinpoint ovulation. In the high-tech method of sex selection, setting the time of ovulation is employed either to harvest eggs for later use or to do a timed medical procedure to help ensure that pregnancy occurs, with a baby of whichever sex you have chosen.

During the first half of your menstrual cycle, called the follicular phase, the levels of estrogen will increase. This helps build the uterine lining so that it can accept a fertilized egg for pregnancy. Simultaneously, a follicle is maturing with the ovum inside, as a result of the production of a hormone known as follicle-stimulating hormone (FSH). As your body has a surge in the luteinizing hormone (LH), known as the LH surge, the egg is released in ovulation.

The second half of your menstrual cycle, after ovulation, is known as the luteal phase. It's called that because it begins with the formation of the corpus luteum, the spot where the egg broke free from the ovary, which will now produce progesterone to sustain a hopeful pregnancy. In this phase the levels of progesterone rise, helping to prepare the uterus for pregnancy. Meanwhile, the egg is on its journey through your reproductive tract awaiting fertilization by sperm.

If the egg is not fertilized, no pregnancy has occurred. Then the progesterone and estrogen fall sharply. This causes your uterine lining to shed itself, restarting the menstrual cycle to day one.

Ovulation

The process of ovulation is a primary part of the menstrual cycle. Ovulation is critical for conception to take place. Without an egg to fertilize, there can be no pregnancy—girl *or* boy. Therefore, if you are fertile, you must be ovulating. If you are not ovulating you will have a confounding fertility issue. Many of the high-tech methods of sex selection can help you overcome this fertility problem, as well as increase the likelihood that you will have a baby of the desired sex.

For most women, ovulation takes place during most menstrual cycles. If a woman isn't ovulating, she's said to have a fertility problem, namely an ovulation disorder. An often-quoted average is that ovulation takes place 14 days prior to the beginning of the *next* menstrual cycle beginning.

Examples:

25-day menstrual cycle — ovulation on day 11
28-day menstrual cycle — ovulation on day 14
31-day menstrual cycle — ovulation on day 17
43-day menstrual cycle — ovulation on day 29

Yet these are only averages, used to show the range of "normal." This range actually extends even beyond what is shown here. And because most methods of sex selection, both high- and low-tech, require that ovulation be pinpointed exactly (to the best of your ability), averages need to be avoided.

Occasionally, during ovulation more than one egg is released. This can mean that a woman will get pregnant with more than one baby, leading to fraternal or dizygotic twins, meaning that two separate eggs were fertilized.

Early division of the single egg-and-sperm union can also cause twins or other multiples. This results in identical or monozygotic twins. In general, the number of multiple births increases when parents use technology to get pregnant. This is true even after using simple, ovulation-inducing medications like Clomid®, though the number certainly rises with the use of in vitro fertilization (IVF). The number of multiples conceived from in vitro fertilization depends largely on the number of embryos returned or placed back inside the uterus at the time of the transfer of the embryos.

Conception

Ovulation occurs and an egg is released from the ovary. The fimbria sway back and forth, trying to move the egg into the fallopian tube. This is where the egg will begin its journey toward the uterus.

The fallopian tube is connected near the top portion, or fundus, of the uterus. It is in the fallopian tube that conception occurs. Once a single sperm penetrates an egg, the race is over. The sex of the baby is determined, but the journey is only just beginning.

From here the fertilized egg will slowly make its way down the fallopian tube. The end location, with luck, will be the uterus, which has built up a nice thick lining to nourish the fertilized egg. Once the egg burrows into that lining, a pregnancy is said to have occurred. By a variety of signals the body will now know it's pregnant, so ovulation and the menstrual cycle will stop for the duration of the pregnancy.

3

Testing and Procedure Basics

The key to almost every method of sex selection is to accurately predict ovulation. This means the ability to predict before-hand, when you ovulate. Knowing the time of ovulation helps you time intercourse to maximize your chances of conception, attempt to do a low-tech sex selection technique, or instead choose a high-tech, more accurate method of selecting for sex.

The basal body temperature (BBT) method of fertility charting is one of the more accurate ways to predict when you ovulate. The beauty of this method is that it's very inexpensive to do and can be done no matter the length of your cycle. Others who believe strongly in this method talk about the additional benefits it provides, by way of knowing your cycle characteristics.

To use this method, you need a thermometer specifically designed as a basal thermometer. This means that it measures your temperature in tenths of degrees. The smaller units of measure are good for detecting slight shifts in your temperature, making you more aware of potential changes in your body. These shifts are key in determining when you ovulate.

A basal thermometer can be purchased at nearly any drugstore. Some high-tech versions are available at many online retailers. For the purposes of conception, nearly any basal thermometer will do, as long as it measures in tenths of degrees, which is what's needed to detect the minute changes in your basal body temperature.

You'll also need something with which to chart. Some families choose the standard paper and pen method of charting. This works very nicely and is low-tech. The chart that many people use will typically be found for free either with the basal thermometer or online. Simply make multiple copies of the chart to use for multiple cycles. I would encourage you to play around with different charts to find one that works the best for you. If you have a chart that's easy for you to use, then you're more likely to use it regularly, and therefore you'll derive more benefits from your charting.

Various computer-based programs are also available online for you to do your fertility charting. Some of them are free and offer handy databases as well as some neat tools, including ovulation prediction based on past cycles, and the like. Certain programs charge a fee, usually under $10 a month for the manufacturer's services, including enhanced charting, more detailed storage of information, and other extras, depending on the website. Many women choose this option because of the ease of sharing and storing information from cycle to cycle.

Taking Charge of Your Fertility, the website inspired by the book, has a unique set of programs online to help you track your menstrual cycles in an attempt to learn more about your cycles, including pinpointing your ovulation. You can find this site at *www.ovusoft.com.* Fertility Friend is another such site; you can view its offerings at *www.fertilityfriend.com.* Many of these sites offer a free trial of use.

A ferning microscope can also be used. This is a small, pocket-sized microscope used to test oral or vaginal secretions. When changes in these secretions occur, a distinct "ferning" pattern can be seen through the microscope. These small microscopes are nice because they're reusable over and over. This can save money in the long run while helping you predict ovulation. They are also a great tool if you are someone who wants to test every hour, which you can easily do with this type of testing device.

The ferning microscopes have been harder to find in retail stores. Many online merchants do offer them for fairly reasonable fees. A quick search online will turn up a couple of brands; one is Maybe Mom®.

The high-tech method that many people use, and that many fertility centers consider the gold standard, is the ovulation prediction kit (OPK), which is sold over the counter in many stores. These kits use urine to predict ovulation by detecting the LH surge in your cycle. You begin testing on a certain day, based on the length of your cycle.

Using an ovulation prediction kit can get expensive if you have a long cycle, because you will need one stick for every day on which you test. An inexpensive testing kit might cost just under $1 per stick.

You can find the fertility monitors in some local stores and at online retailers. I would encourage you to shop around for the best price, as the differences in cost vary widely. You'll also want to price check where to buy replacement monitoring strips for your monitor. Sometimes you can find great deals if you buy in bulk.

Each method of ovulation detection has its benefits and risks. You'll find that many women use a combination of many, or even all, of these detection methods. This allows them to feel that they have covered all their bases and increases their confidence in conception.

Infertility

The inability to conceive a child after at least one year of well-timed, unprotected intercourse is called infertility. It affects several million couples a year in the United States alone. It's said that 1 of every 7 couples is experiencing infertility. The problems that cause infertility can originate with the man, with the woman, or with both—or can just be unexplained.

Testing by a reproductive endocrinologist or fertility specialist can help you define what the problem is with conception. It might be related to ovulation or the menstrual cycle. It could be with sperm or egg production. It may be a hormone issue. There are many reasons that you may be infertile. The good news is that many treatments are available to help couples experiencing infertility.

Some of these assisted reproductive technologies are also what are beginning to be used to help couples select babies for their sex for reasons other than medical need. A number of these techniques were found accidentally, like the fact that more females are born after a couple of uses of Clomid® to induce ovulation. Then there are techniques like MicroSort® and preimplantation genetic diagnosis (PGD), both of which have been used for sex selection.

Medical Procedures You May Have to Undergo for Sex Selection

When you begin to look into sex selection, you will find that there is nothing natural or easy about any of the methods, whether high- or low-tech. In general, though, the higher tech you go, the more medically intensive the whole process becomes. This can lead to an increase in the number of medical procedures that you and your partner need to understand or undergo, or both.

The basics of those procedures are listed in this chapter. While each form of sex selection may use only one or two of these procedures, knowing about each will prevent surprise and help prevent miscommunication or inappropriate expectations between you and your practitioners.

Intrauterine Insemination (IUI)

The intrauterine insemination, or IUI, is used as a way to bypass intercourse as a means to achieve pregnancy. For some, the procedure is chosen to deal with fertility issues, such as certain forms of male factor infertility, hostile cervical mucous, the use of donor sperm, or other needs. But intrauterine insemination is often employed in sex selection as well.

The reason for the popularity of intrauterine insemination in sex selection is that most of the other techniques typically done on a semen sample will leave the sample with such a small number of sperm that the chances of pregnancy are relatively low. The goal of using intrauterine insemination is to increase pregnancy rates when doing sex selection.

Before an intrauterine insemination is scheduled, you must first pinpoint your date of ovulation very precisely. Once that date is found, you will call the fertility center to schedule your intrauterine insemination. This appointment is usually within 24 hours of your call. Sometimes it can be as soon as the same day, depending on when you tested for ovulation and what type of ovulation testing you were doing.

You may be asked to come to the fertility center early to provide a semen sample before your intrauterine insemination. This will not be necessary if you are using donor sperm (which will already have been tested) or sperm that has been previously processed and is stored at the fertility center.

The actual intrauterine insemination is a very quick procedure, much like a Pap smear. You will be asked to undress at least from the waist down. The practitioner, who may be a doctor or a nurse, will then insert a small plastic catheter into your uterus to deliver the specially prepared specimen of sperm. Some women report that it feels a bit strange, though most say that other than a slight pinch as the catheter goes through the cervix it's not bad at all. All in all, the business portion of the intrauterine insemination takes only a minute or two.

Some couples try to make this a special time, since they hope it will result in the conception of their child. One couple I know were really upset at their doctor, who just barged in, very businesslike, without even a hello. After making him stop to talk to them about their ideas, they brought him to realize that they needed him to go more slowly. He actually wound up helping make them more comfortable by talking them through a nice relaxation exercise just before he did the actual intrauterine insemination.

When this couple got pregnant from that intrauterine insemination, they felt that it was the relaxation as well as the doctor's change in attitude that helped them. While there are no studies to say that either way is best, it's true that any success is at least partially about the experience. Either way, it worked for the couple, and they were thrilled.

Consider what might make your insemination feel special to you. It may be a relaxation exercise, or it may be music or any other personal touch. Many women bring good luck charms on their visit. Several of the mothers interviewed said that they brought something with them that represented the sex of the baby they were hoping for. Explains one such mother:

> "When I woke up that morning I chose these horrendous pink socks because they screamed 'girl!' to me."

Depending on the protocol of your fertility center, you may be asked to wait for a bit before leaving after your insemination. Other centers have you immediately leave as soon as you are ready. Discuss this protocol before your intrauterine insemination so that you can make an informed decision. This also prevents surprises on the day of your treatment.

Hardly any home care is needed after your insemination. One mother said she wished she had known earlier that sometimes she would feel a bit of fluid seeping out. Her main concern was that the semen was leaking out and therefore that her chances of pregnancy were falling. This is not true. In fact it is common to feel "wet" after an intrauterine insemination.

Out of superstition, there are certain things that some women do, or don't do, at home. Sometimes women will resort to doing rituals or taking actions that they know they did in other successful cycles. Others will be overcautious, just in case. The best advice is to just "act pregnant," the way any other mother trying to conceive should be doing at that point in her cycle. This can include avoiding potentially hazardous substances like alcohol and should include measures like taking prenatal vitamin supplements to protect your future baby.

You will be instructed to report to the fertility center for a pregnancy test in 14 days. Some centers allow you to do local or home testing if you live far away from the clinic. This is also something that should be discussed early on in your care. It is unwise to try to take home pregnancy tests before this time. While some claim to be sensitive enough to detect pregnancy prior to the 14-day window, it doesn't pick up every pregnancy this early. This can lead to emotional turmoil when it's not necessary. Unfortunately, pregnancy tests can't tell you the exact amount of hCG in your urine; they can only tell you if your hCG level is above or below the threshold of that specific pregnancy test.

Collecting a Semen Sample

Semen samples are usually required for virtually all methods of high-tech sex selection. The reason is that it is in the sperm where the sex of the baby is determined. By collecting the semen, you can separate the sperm from the seminal fluid and work your sex selection magic on it—of whatever variety you choose.

Many fertility centers that also offer sex selection prefer to do a basic semen analysis before proceeding with sex selection techniques. This is by no means done by all of them. Many find that the small cost of a semen analysis is worth it to a couple, to save them from spending a lot of time, effort, and money only to discover that there is an issue of male factor fertility at hand.

After the semen sample is delivered, doctors and lab directors look for problems with the sample, such as a low sperm count, decreased mobility of the sperm, abnormal shapes or other malformations, and other issues that might be noticeable in a sample. They can only do this by looking at actually ejaculated semen.

Since this test is relatively easy and painless, most couples do not object to the testing. It consists of collecting ejaculate—that is, the fluid that leaves the penis during male orgasm—in a container to look at it under a microscope.

Most practitioners prefer that the male partner produce the semen sample at their offices, though alternatives are sometimes available. Some practitioners will agree to collection done at home, if it can be brought to the center within a specified amount of time, usually about an hour. (As always, check with your fertility center for their rules.) Some women report, romantically, that carrying the container of semen in their bra helps keep it warm for the drive to the clinic.

You will collect the specimen in a container made specifically for this purpose. It can be marked with your name or an identifying number to prevent lab mix-ups. This rigor usually allays most couples' fears of its getting lost. Occasionally you can find a practitioner willing to let you use a special collection condom made from polyurethane. If you use a condom with intercourse to collect the ejaculate, you must use the special collection condom. Regular condoms, even without spermicide, may negatively affect the semen analysis.

Preparation for the semen analysis usually involves a short period of abstinence from all forms of ejaculation. This means no sexual intercourse, no masturbation, no oral or anal sex, for at least three days before the test. The purpose is to get the best sample possible for the analysis.

When collecting the sample, you can allow your partner to collect it. The infamous collection rooms and their "reading" material do exist to help set a sexy mood. Many women choose to help their partners in the collection process. The collection process can be a team effort, through masturbation with a partner. It is wise to point out, though, that oral sex is not a good collection option because of the damage that saliva can do to the semen sample.

Once your doctor has the sample, she or he will perform numerous tests on it. These include the volume of the total sample, the number of sperm present, how mobile (or "motile") the sperm are, the contents of the seminal fluid, and the size and shape of the sperm. If these are all found to be within normal limits, you will be cleared for further steps toward sex selection.

Sometimes, however, these factors are outside the normal limits. This can happen for lots of reasons, including recent sexual activity or illness. Even something like alcohol or tobacco can influence the semen sample. Your doctor might recommend ways to boost the outcomes. Something simple might be suggested, like taking vitamins, avoiding smoking, or even wearing boxers, depending on the outcome of the testing.

While the cost of a semen analysis will vary from lab to lab, it typically runs less than $250 per sample. Some labs include this test with the cost of the sex selection process, while others do not. Be sure to ask up front if this lab work is included. Some insurance companies or plans will cover the cost of the semen analysis, even if they happen not to cover other costs of sex selection or fertility treatments.

Ovulation Prediction

Predicting ovulation is the heart and soul of trying to get pregnant. Whether you choose to do anything about the journey in sex selection, if you have chosen to have a baby, you will most likely need to know something about ovulation. The most common fact to know is when you ovulate, because it is ovulation that signals when you're most likely to be fertile and therefore able to conceive.

Tracking ovulation is an ancient skill, passed along from mother to daughter. Today's methods of tracking ovulation can incorporate a bit of the old tried-and-true methods as well as some high-tech, hormonally based methods. Many women choose to use a combination of methods, either because of the added assurance or because of cost.

Ovulation generally occurs once in each cycle for a woman who is having normal menstrual periods. It is not unheard of to have a cycle when she does not ovulate at all, though most women will be oblivious to it. Problems associated with charting, or discovered by it, may indicate a lower level of fertility, which might change your method of choice for sex selection.

Signs that you might be having trouble ovulating can include:

- Irregular cycles
- Heavy bleeding
- Lack of cycles

If you experience any of these symptoms, you should report them to your practitioner. Often therapies are available that can help you overcome ovulation problems fairly easily. Only your reproductive endocrinologist can help, however.

Basal Body Temperature (BBT) Methods

Detecting ovulation through the basal body temperature method is also known as fertility charting. It has been used for a very long time, both to increase the odds of conception and to prevent conception.

With this method, as briefly described earlier, you will use a basal thermometer, a special thermometer that measures in tenths of degrees, to take your temperature first thing every morning.

This should be the first thing you do before you even get out of bed. You can take your temperature orally, vaginally, or rectally. The key is to always use the same method, at the same time and in the same location each day.

You will do this charting on a special piece of graph paper. You will note anything that changes your temperature—whether it be a late night out, more physical activity than normal, illness, and, of course, ovulation. Because not every woman ovulates within the 14-day window that many so-called conception experts talk about, charting will help you figure out when you ovulate, though it can take several cycles to do so accurately.

Your best clue as to when ovulation occurs will be a shift of at least four-tenths of a degree over temperatures in the first phase of your chart. Some women will notice a drop and then the rise, indicating that ovulation has occurred. The dip may be your cue to predict ovulation, should you need to know before you ovulate. Once you ovulate you're considered to be in the luteal phase of your menstrual cycle.

The continuation of charting is also helpful at detecting pregnancy. Toward the end of the luteal phase of the cycle, your temperatures will normally drop. If a pregnancy has occurred, the hormone levels are such that the temperatures stay up, as indicated by the basal body temperatures.

One of the many benefits of using the basal body temperature charting is that it is inexpensive. Other than the purchase of the thermometer for around $10, there is no other forced investment. Some women choose to use books to help them learn the finer points of charting as well as the symptoms that their body shares as cues to fertility.

Many women also prefer to chart as a way to stay in control. One of them, April, shared with me that she needed to know when she ovulated anyway, so:

"Why not use what I already had for free? It gave me a sense of control, even if it was all in my head."

Many women echo what April has expressed. They gain a sense of control from recording their temperature every morning and then using that data to interpret bodily symptoms. They also find it nice when it confirms what other methods are telling them about their cycle. For some women this is the only control that they feel that they have in the whole process.

"I'm a control freak, I'll admit it. But I couldn't give everything up," relates one anxious mother-to-be. "Taking my temperature gave me something to do and more importantly something to write down every morning. It kept me focused and sane during the insanity of the whole process."

Ovulation Prediction Kits (OPKs)

Stores everywhere sell ovulation prediction kits (OPKs). They are not at all difficult to find. These work by detecting the luteinizing hormone (LH) that is released at the time of ovulation. These kits use your urine to screen for LH.

You may need a number of test kits throughout your cycle. How many kits you need will depend on how long your cycle is typically, and when you ovulate. Some women use about 7 monitoring strips or tests in a cycle, while others use 20 for very long or odd cycles.

The ovulation prediction kits found in drugstores can be quite costly, at several dollars a kit. Many online stores sell ovulation prediction kits in bulk. While the quality is the same, these kits have less packaging. For some women, the easier-to-use, larger-packaging kit is worth the extra money. Overall, though, there is typically not much difference in the kits, what they test, or how sensitive they are.

Depending on the lab or practitioner with which you are doing your sex selection, you may be requested to use a certain brand of kit or method. Be sure to find this out before buying any product in bulk. You will also want to check for expiration dates with tests of this nature. And it's important to point out that these tests are one-time-use only.

Other Monitors Available for Ovulation Detection

Two additional types of fertility monitors are worth mentioning. One is something simple, inexpensive, and reusable: a simple lens or microscope that uses your saliva or vaginal secretions to detect ovulation. The other is a fertility monitor that measures hormone output from your urine.

The microscope, or lens, is the size of a lipstick container. It's that small, and it typically costs about $40. You place a drop of saliva or vaginal secretion on the lens and slide it back into the case. When you look through one end of this small device you see the pattern the fluid makes when dry. If the pattern looks like a fern plant, you're ovulating.

The benefits of using one are that it provides highly accurate and inexpensive results. Because it can be reused thousands of times, you could literally check every hour for ovulation without added cost. This is great for women who are obsessing about getting pregnant and need something to do instead of thinking about it. It's easy to use, and most women really like it.

The other fertility monitors are actually fairly expensive to purchase. They also require that you purchase monitoring sticks, for about a dollar a day. This price may come down if you purchase the test sticks in bulk.

While you can't use this machine whenever you want, it will record data, predict ovulation, and not only tell you when to test for ovulation based on your cycle, but also give you a cycle count. This machine is a bit fancier and many women like its high-tech feel, though it is not good for use with long cycles.

In Vitro Fertilization (IVF)

Who Is the Typical Patient for In Vitro Fertilization?

An infertility patient is typically someone who has been trying for a long time to get pregnant. She may or may not have spent years using low-tech methods to conceive, though chances are that she's spent some time doing "simple" infertility treatments like Clomid® and intrauterine inseminations (IUI).

Some of the women and men who suffer from infertility are also interested in doing sex selection. This may be for medical reasons, such as reducing the likelihood of a single gene defect, or for other, more personal reasons.

In vitro fertilization can be a frightening process. If you prepare yourself ahead of time with information about the process, you likely won't feel so alarmed. One of the most important things to know is that you are not alone. Many families have done "in vitro" before you, and have had great successes and gone on to bear babies. There's no reason you can't be one of those families.

One of the most frustrating things about in vitro fertilization is that it's a lengthy process. Rather than speaking of days, we're talking weeks and months of scheduling, which is the first step in the process. Deciding when to start an IVF cycle will depend on your menstrual cycle, the availability of your clinic and their personnel, and your availability. You may also need to look at other factors.

Before doing a cycle of in vitro fertilization, you will need to do some precycle lab work and testing. Commonly, the first step is lab work. This will include blood work, but it may also include a semen analysis, depending on when the last one was done. You may be required to do other testing too.

Many fertility centers will want you to do, at the very least, an ultrasound of the uterus. This is done to screen for uterine anomalies that might prevent your from getting pregnant. Other centers require that you have a hysterosalpingogram (HSG), in which dye is injected into your cervix while X-rays are taken. This test allows doctors to check both for uterine anomalies and for any blockages in the fallopian tubes.

While each treatment cycle will vary somewhat, in length and in the exact dosages of medications, there will be some similarities, depending on your normal fertility, your body's response to the medications, and the criteria of your clinic. In most centers you'll be assigned to a treatment coordinator or other case manager, who will be specifically available to answer your questions or find answers for you. This can be a great thing, because it keeps you from getting the runaround, which can be frustrating when it's 10 o'clock at night and you're not sure which medication to get, or how.

The precycle warmup typically includes a prescribed month or more of birth control pills. The reason behind using oral contraceptives is that they are an inexpensive, effective way to quiet your ovaries and keep your female hormones in check. By doing this, your doctor and staff have a better chance of controlling your cycle with other medications to come. This can be baffling for some families. Cried one mother:

> "I wanted to be pregnant! So why would I do birth control pills? I totally understood on a mental level, but every night as I'd choke down a pill, it was very hard to understand emotionally."

Then there are others who are simply glad to be started on their journey, no matter what leg of it they're on. This can be particularly true if the wait has been long, for whatever reason. One mother explained that she felt the process was all about her attitude and perspective, pointing out that she could choose to be excited as a valid option.

Getting Started

If you are using in vitro fertilization, regardless of your fertility status, your medication protocol may—and most likely *will*—be quite different from that of other women you know, even at similar stages and ages. This is not at all unusual. You should never attempt to change the amounts of medication you take, because doing so may add risks and possibly even cause your cycle to be cancelled. You may also find, if you discuss your cycle protocol with others, that your fertility center does things radically differently than others. This is common and not a cause for concern or alarm. Remember that in getting pregnant, there are many ways to achieve successful results. If you have questions or concerns, you should take them to your reproductive endocrinologist and get an explanation.

After the oral contraceptive phase of the cycle, most commonly a gonadotropin-releasing hormone agonist (GnRH agonist) will be employed to down-regulate the patient's own hormonal control of ovulation and allow the doctor better control of the ovarian stimulation portion of the cycle. The most common such agonist is called Lupron®, given as a subcutaneous (sub q) injection. The Lupron® fools the body into believing that estrogen levels are very low. Cetratide®, a GnRH antagonist, may also be used for this same purpose.

GnRH agonists can be started either at the end of the previous cycle or in the earliest portion of the current cycle.

The next step in the cycle of in vitro fertilization is to stimulate the ovaries. This is done with gonadotropins, medications that cause your ovaries to respond by producing oocytes (eggs). A typical length of gonadotropin treatment will be between 6 and 12 days, depending on many factors. The type of gonadotropin most often used is follicle-stimulating hormone (FSH). Many different brand names and variations of FSH may be used. Two of the more commonly prescribed are Gonal-f® (follitropin alpha) and Follistim® (follitropin beta). Occasionally your doctor may also use a bit of a second gonadotropin, luteinizing hormone (LH), in combination with FSH. Most of these medications are given subcutaneously, but a few can or must be administered intramuscularly (IM). Other names you may hear include Bravelle™ (FSH/LH, IM, or SQ), Repronex® (FSH/LH, IM, or SQ), and Menopur® (FSH/LH, SQ).

Since you will be giving yourself these medications in shot form, you will need to learn how to give yourself injections. Self-injecting medications can be very frightening, particularly at first. Some women find friends to help, or teach their partners to inject the medications.

You will be carefully monitored throughout your cycle. This monitoring will include both laboratory blood work and frequent ultrasound screenings. By carefully following your hormonal levels and the size and number of follicles, doctors can determine when to trigger the final development of the eggs, when to schedule the retrieval, whether the uterine lining is optimal for return of fertilized pre-embryos, and also whether you are at risk for developing ovarian hyperstimulation syndrome. If OHSS develops, the cycle may be cancelled.

Ovarian hyperstimulation syndrome is a condition in which the ovaries enlarge so drastically and quickly that they leak fluid into your abdomen and cause multiple symptoms. These can include nausea, dizziness, weight gain, shortness of breath, and others. Few women will need to treat ovarian hyperstimulation syndrome, though occasionally a woman will need to be hospitalized. Most symptoms go away completely, though slowly, usually within 10 weeks. This phenomenon occurs in less than 1 percent of the patients undergoing ovarian stimulation. If a patient is likely to develop ovarian hyperstimulation syndrome, signs and symptoms usually appear within 3 and 6 days after ovulation. In addition to physical discomfort and medical risks for the patient, ovarian hyperstimulation syndrome can cause a cycle to have to be cancelled. For both these reasons, it is something that your practitioner will try to prevent.

During the ultrasound screenings, your physician and his or her staff are trying to assess when you have reached the maximum number of oocytes, the optimum size of the follicles, and the condition of the uterine lining. During the menstrual cycle, the uterine lining grows to prepare to receive a fertilized egg. The uterine lining should usually be from 8 to 16 millimeters thick to be considered appropriate for implantation.

While it is tempting to get caught up with pondering the right number of follicles, there is also the fact that in oocytes, size does matter. Typically the goal is to have the follicles be between 16 and 23 millimeters in length. Once your doctor believes yours have reached this point, you will receive another injection.

This injection is of human chorionic gonadotropin (hCG). Because hCG is chemically similar to the luteinizing hormone that signals the follicles on the ovaries to ripen and release their oocytes, it is used now to help ready the follicles for the egg retrieval. So the hCG injection is used as a form of luteinizing hormone (LH) surge. The hCG injection is given about 36 hours before the retrieval is to be done.

The more commonly prescribed hCG medications are Pregnyl®, Profasi®, and Novarel®. These medications are given as an intramuscular injection.

Other medications may be used in any given cycle of in vitro fertilization. During the course of your routine blood work, if it is found that your estrogen levels need supplementing, you may be prescribed an estrogen patch. Some women also have other disorders that require treatment during the cycle, such as treatment with baby aspirin for blood clotting disorders. Estrogen levels may be supplemented with patches or baby aspirin, and/or vaginal Viagra® may be used to increase blood flow to the uterine lining. Patients with specific fertility disorders such as PCOS (polycystic ovarian syndrome) may be asked to use medications such as Metformin® as part of their regimen. You should not be surprised if your doctor or clinic suggests additional medications for your cycles, based on your underlying health conditions and their professional experiences and protocols.

Other GnRH agonists include:

- Leuprolide (Lupron®, Eligard®)
- Buserelin (Suprefact®, Suprecor®)
- Nafarelin (Synarel®)
- Historelin
- Goserelin (Zoladex®)
- Deslorelin

All of these may be administered as sub q injections or, in the case of Synarel®, as a nasal spray.

Egg Retrieval and Sperm Collection

Once you have your appointed time for your retrieval and your hCG shot is done, you simply show up at the fertility center to await the removal of the oocytes from the follicles. The sperm collection will need to have happened before and be brought with you, or be scheduled to happen that same day at the clinic. If you're using donor sperm, the sample will already be waiting for you at the fertility clinic; you won't need to do anything special for this service.

If you are doing MicroSort® at the Huntington Reproductive Center or at Genetics & IVF Institute, the sperm sample will be made earlier that day and processed for your choice of an XSORT® or YSORT®. If you are using a local collaborator, the sorted sample has been prepared before-hand and shipped, frozen, to your collaborating physician. The lab at your fertility center will then be responsible for preparing the specimen.

On the day of your retrieval, you will be asked not to eat or drink for at least 6 to 8 hours before your appointment. This is for your safety, because staff will typically be using some form of sedation or general anesthesia to help ease the physical and mental discomfort of the egg retrieval.

The vast majority of women will have the egg retrieval done as a vaginal ultrasound-guided procedure. The ultrasound is used to guide the needle to find the oocytes within the follicles. Each follicle is aspirated and the contents are handed to the embryologist in the lab. This procedure usually takes 30 minutes, though it may take longer in some women.

Occasionally a woman will require an abdominal retrieval, which is done through an incision in the abdomen with a laparoscope. You may need this if you've had certain surgeries or have certain anatomical issues that prevent the reproductive endocrinologist from doing the retrieval vaginally. You may know this ahead of time, or at least you may know that there's a possibility of your needing an abdominal retrieval.

By the time you are resting in the recovery room, the doctor is usually able to tell you how many eggs were retrieved. The number depends on your age, the medications you are taking, and your body's response. Your eggs will be inspected by the embryologist and her or his staff.

Progesterone Injections

Normally, the corpus luteum would produce progesterone to help build and sustain the uterine lining. Now you need help with that, because the corpus luteum might be compromised in the process of retrieving your eggs.

Progesterone can be given by injection, as an oral medication (Prometrium®), as a vaginal gel, or even as a vaginal suppository. In most cases treatment is started on the day of your retrieval though procedures in your clinic may vary.

Your Embryos

From the time when your reproductive endocrinologist hands over the follicular fluids, the embryologist goes to work. First he or she will examine all the fluid that has been collected, removing any oocytes (eggs) that are found. Either the eggs and the prepared sperm are placed together in the petri dish to progress to fertilization, or an intracytoplasmic sperm injection (ICSI) is used to fertilize the eggs. (ICSI is often used in cases of preimplantation genetic diagnosis to ensure that only one sperm joins with an egg.) The embryologist will usually call you to give you updates. The news might include how many eggs were retrieved, how many actually got fertilized, and how they are all doing.

The fertilized eggs, or pre-embryos, are graded based on the number and appearance of cells. By day two, 48 hours after the retrieval, it is preferred that you have four celled embryos. At day three, 72 hours after retrieval, seven or more cells are preferred. Ideally, the cells will be similar in size and shape, with little fragmentation around them. You will usually not have a cell graded until day two.

It's generally believed that the higher grade the cell, the more likely a pregnancy is to occur. The problem is that perfectly fine-looking embryos can fail to implant. It is also true that very low-grade embryos can be transferred and still successfully implant. This is where preimplantation genetic diagnosis has made huge strides. Fertility specialists and embryologists have learned that cells can look perfect but still be of poor quality, or can carry genetic imperfections like severe diseases. Simply put, there is no way to look at an embryo and be able to predict its quality perfectly. This is where other genetic testing, like preimplantation genetic diagnosis, comes into play.

How Many Embryos Should Be Put Back?

One of the biggest decisions regarding in vitro fertilization is the number of embryos to be put back in the embryo transfer. This number has been argued back and forth by many entities. Some physicians believe that more is better, but in general that has not been borne out by the science. This controversy is ongoing.

The American Society for Reproductive Medicine (ASRM) published guidelines in October 2006 outlining the number of embryos that should be transferred. The number of embryos, it said, independent of maternal or donor age, should depend on the prognosis of pregnancy. A pregnancy is more likely in the first cycle of in vitro fertilization and if the quality of the embryos is high. The guidelines also recommend looking at the number of good-quality embryos available for cryopreservation.

For a woman who is under 35 years of age with a good prognosis for pregnancy, it is recommended that only one embryo be placed into the uterus. If the prognosis is not as favorable, no more than two embryos (cleavage-stage or blastocyst) should be transferred. Two cleavage-stage embryos are also recommended for women between age 35 and 37 with favorable pregnancy prognoses. ASRM recommends that three embryos be the limit for women of the same age range but with less-positive prognoses. For women in the 35–37 age group, no more than two blastocysts should be returned to the uterus. For women between 38 and 40 years old, ASRM suggests that three cleavage-stage embryos or two blastocysts are sufficient if the prognosis is favorable, but up to four cleavage-stage embryos or three blastocysts can be considered.

Women over 40 are not eligible for the MicroSort® procedure, but may do preimplantation genetic diagnosis alone. ASRM says that for that age five is the maximum number of cleavage-stage embryos, or three blastocysts, to place in the uterus. Its guidelines also urge women to make decisions based on their specific situation, including their cycle history. For example, even if you're younger than 40, but have a history of two or more failed cycles, you might consider adding more embryos.

If you are using donor eggs for your in vitro fertilization cycle, you should look at the age of the donor, rather than your own age. So if you're over 40 and your donor is 28, you would use the guidelines for embryo return based on the age of 28, meaning that fewer embryos are placed back into the uterus.

It is wise, however, to point out here that part of the reason for implantation failure actually lies in the quality of the embryos. When preimplantation genetic diagnosis is used, the pregnancy rates, even for women in their 40s, are very similar to that of younger women. This is supposedly because of the higher-quality embryos. Therefore, one could surmise that high-quality embryos, together with a large quantity of them, are more likely to result in a multiple pregnancy, even in a woman over 40.

The rationale for taking a moderated approach to embryo return is to reduce the risks of multiple pregnancy. Multiples pregnancies are more likely using this technology, plus risks are inherent with multiples, both to the mother and to her babies. For the babies, these risks include pregnancy loss, preterm labor, premature birth, low birth weight, and others. Mothers of multiples often have problems associated with bed rest, preterm labor, high blood pressure, gestational diabetes, and other severe complications of pregnancy and birth.

Embryo Transfer

The embryo transfer will happen two to five days after the retrieval. Which day will depend on whether your clinic has a policy in place, and whether you did preimplantation genetic diagnosis. It may also depend on the quality of your embryos. For most women, the embryo transfer is seen as anticlimactic. One mother recalls:

> "I remember lying there while the nurse transferred the two embryos back into me. It was like 'That's it?' After everything else being such a big deal, the 'conception,' at least my part, was pretty low key."

A doctor or a nurse may do the vaginal exam–like procedure. In some clinics a resident in obstetrics and gynecology or a Fellow in reproductive endocrinology may also do your embryo transfer. Be sure to ask, if you aren't told, who will attend your transfer. Many parents say that they wish they had done something special during the transfer to mark the solemnity and joy of such a moment, such as play favorite music, pray, or meditate.

The actual transfer of the embryos by catheter into your uterus is a gentle one. Your practitioner will delicately release the embryos into the uterus. This will give them the best chance to gently stick to the surface, then burrow or implant into the uterine lining. The whole procedure may also be done with ultrasound guidance. (If you do a transfer with ultrasound, you might experience discomfort due to a need for a full bladder.) After the procedure, the embryologist will also ensure that there are no embryos left in the catheter.

Other Forms of Embryo Transfer

For a variety of reasons, a standard embryo transfer may not be your best option for getting pregnant. Luckily, there are multiple ways to get your embryos or genetic materials back inside your body where they can go to work to make a baby.

Gamete intrafallopian transfer (GIFT) is used to help some couples get pregnant, though technically it is not an embryo transfer. If a couple also desires sex selection, then MicroSort® is also necessary. Here, while the cycle looks very similar to in vitro fertilization, the eggs and sorted sperm are placed inside the woman's fallopian tube via a laparoscope. This allows conception to take place naturally inside the fallopian tube. The fertilized egg will then travel down through the fallopian tube and, with any luck, implant inside the uterus. Gamete intrafallopian transfer has very high success rates, particularly if the couple typically has not done well with in vitro fertilization.

Zygote intrafallopian transfer (ZIFT) is another option for embryo transfers. It is also known as tubal embryo transfer (TET). In these methods, about one day after the retrieval the embryos are placed into the fallopian tube using a minor surgical procedure. While this does pose a higher risk of multiples pregnancy, you will know how many eggs actually fertilized, unlike with the gamete intrafallopian transfer.

Another benefit of zygote intrafallopian transfer is the ability to use some forms of sex selection. These include the use of intracytoplasmic sperm injection (ICSI). This can dramatically increase the chances that your sex selection will be accurate.

The Two-Week Wait

And now the hard part: you sit back and wait. Your recovery from the egg retrieval is usually straightforward. You may want to take it easy for a day or two. Many women feel slightly "out of it" from the anesthesia. You will be advised to stay home from work and not to drive a car or make important decisions for 24 to 48 hours.

Your fertility clinic should call you with updates on what is going on in the embryology lab. At specific intervals, you may receive reports on fertilization and cell division. Ask before you leave the clinic when you can expect to hear news.

The hardest part of the two-week wait is the emotional strain and turmoil. Mary remembers this about her wait:

> "It was not a fun thing, but you're obsessed. Doing pregnancy tests at day 10.... It's a roller coaster of emotions."

Many women need a lot of support during this fallow period. This is a great time to call on that support system that you have built. A number of online fertility forums are filled with messages from women obsessing about their two-week waits, and many of them receive knowing advice and compassionate support from others who have been there, or who will be there in the near future. This support can be invaluable while your mind goes back and forth with all the possibilities and craziness of waiting for the pregnancy test.

Many women spend this time trying to busy themselves with anything other than obsessing over the outcome of the pregnancy test. Some women said that they did their best not to think about it at all, while others took a more moderate approach and tried to limit the time they spent dwelling on pregnancy in general. The rule here is that you should do whatever mentally and emotionally gets you through this time period, whether it be ignoring every symptom and sign of pregnancy or totally wallowing in it.

Pregnancy Testing

Most fertility centers recommend that you do *not* test for pregnancy until day 14 after your retrieval. With the advent of super-sensitive pregnancy tests, many women do self-testing much earlier.

Pregnancy tests detect human chorionic gonadotropin in either urine or blood. The home pregnancy tests (HPTs) are done with urine. They are readily available and fairly inexpensive, particularly if you buy them in bulk—which is often preferred by women undergoing infertility or sex selection treatments. But this means a lot more of a roller coaster ride because of the testing. One woman, C. K., recalls:

> "I remember testing from day 10 and debating whether or not that negative test was really negative. Was it negative because I wasn't pregnant? Or was it negative because it was simply too early to test?"

Many women promise themselves that they will get off the roller coaster of pregnancy testing, but it simply calls to them, alluringly. April says, laughing:

> "We have this technology. Why not use it? Even if it's driving us crazy. It at least kept me occupied."

Many fertility centers will want to do their own testing to confirm a pregnancy. This is typically a blood test. The blood tests also have the ability to give a specific number of human chorionic gonadotropin (hCG). By measuring this number repeatedly, the reproductive endocrinologist has a better idea of the health of the pregnancy. In a healthy one, the hCG will about double every 48 hours in early pregnancy. Following this is as simple as doing several blood tests. These blood tests can be done wherever you live, so it's not necessary to go to the location of your center, if you traveled for treatment.

The overwhelming feelings of joy and excitement that accompany a positive pregnancy test are indescribable. Many women simply break down in tears of happiness. But now they start a different wait—the wait to see if the sex selection was right on the money.

Typically you'll be seen for an ultrasound test around the sixth week of pregnancy, as calculated from the cycle day one. The test will look for a heartbeat and count the number of babies in the uterus. Another scan may be done around weeks 10 through 12 just for reassurance that all is well. Then you are usually released to your normal midwife or obstetrician. Follow-up care typically consists of only recording the pregnancy outcome and revealing the sex of the baby.

If the test is negative, there is no need to say that the parents feel awful. Their hopes and dreams are shattered in a few moments. Some couples immediately know that they will do another cycle, while others feel so raw and fragile from the current cycle that they can't imagine doing that to themselves again.

Donor Cycles

If you will be using an egg donor, the process for the donor will be much the same as outlined above. The big difference is that your body will be synchronized to the donor's cycle to prepare for the embryo transfer.

Cycle synchronization is done by using a variety of medications. This cycle synchronization prepares your uterus to accept the eggs precisely when they're ready. If that doesn't happen, then the cycle is completely wasted and will have to be scrapped and restarted, which is costly for everyone, mentally and physically, not to mention financially.

Egg donor cycles may actually result in higher pregnancy rates for women who have been suffering from certain forms of infertility. There has been a lot of news about the use of egg donors. Their use in cases involving sex selection, however, has not been covered widely.

PART TWO

Low-Tech Methods of Sex Selection

4

Low-Tech Sex Selection

Timing intercourse. Avoidance of orgasm. Multiple orgasms. Lime douches. What do all these have in common? Believe it or not, they are all used as low-tech methods to try to influence the sex of your baby-to-be.

These tried-but-*not*-true-methods of sex selection appeal to the masses for multiple reasons. For one, they are completely under the couple's control. You do not need to go and explain yourself to anyone while sitting in a doctor's office, half dressed. All potentially embarrassing or personal details are kept within the couple, rather than being told to a medical professional or the couple's family. This appeals to many couples, for obvious reasons.

Low-tech methods of sex selection also appeal to a couple's sense of morality. There is nothing being harmed—no left-over embryos, no sperm that is "wasted," nothing discarded purely for the sex of the embryos. This makes low-tech methods, despite their low probability of success, easier to accept for many couples.

These methods often can be used at little or no cost, but remember: there is no such thing as a free lunch. Simply because they are much lower in cost does not mean that they are actually worth what little they *do* cost. These methods tend to have lower success rates than the high-tech methods of sex selection currently available. So remember, you get what you pay for, even when it comes to sex selection.

While you're reading this chapter, many couples are shelling out hundreds of dollars to charlatans and fly-by-night businesses that promise dreams they can't fulfill. A quick search on the Web for the names of businesses turns up no shortage of people willing to take your money to help you get the sex of the baby you want, even if what they're selling is a dream and is not likely to help you achieve any form of sex selection.

Several of these businesses have had complaints filed against them for failure to provide the services or promises that they purport to be able to deliver. So even if an affordable-price guarantee is made, you need to ask yourself which you'd rather have—the money, or the baby of your dreams?

Many couples will wind up spending thousands of dollars over the course of many years and multiple attempts. Some of these attempts fail to produce a pregnancy, let alone the baby of a specific sex. In fact, after many women and their partners give up, they have likely spent enough money that they could have done a cycle or two of a high-tech sex selection method that would have been far more likely to pay off in terms of successful pregnancies, as well as in a baby of the desired sex.

While detractors of each of these methods obviously exist, those who are successful with them are only too happy to sing their praises. Yet without science and research behind it, we can only assume that in these instances the 50/50 odds of boy or girl fell into the court that they chose. Coincidence, or something done right? No one currently knows.

Positioning

The use of sexual positioning for intercourse is an oft-quoted method of selecting the sex of your baby. You may even have heard some of these old wives' tales. If a woman gets on top during sex, one theory goes, the baby will be a boy. Or the sex of your baby depends on which side you are lying on for conception.

In ancient days, rumors were even floated that women had two uteri, as did the recently autopsied animals in the fledgling doctor's chambers. This led to the belief that there was a girl uterus and a boy uterus. The theory was logical: two uteri, two testicles, two sexes. But this inaccurate view of the male and female anatomy led to a wild gymnastics match to try to line up the "boy" ovary with the "boy" testicle in an effort to have a baby boy.

The problem here, as with most old wives' tales, is that things are often reversed, depending on whom you talk to. Take Mandy's family, for example. They always used the missionary position for sexual intercourse and they always had girls—three of them, to be exact. So Mandy decided to take matters into her own hands and climbed on top one fateful night—all in the hopes of having a boy. Nine months later, her dreams came true. Now, of course, she swears by low-tech methods of sex selection, particularly positioning.

For every Mandy, there is another family who will tell you that specific positions don't matter. It's all the luck of the draw. They point to photographs of their much-loved children, who happen to be of the opposite sex from what the parents were trying for when using a specific position.

Positioning for sex selection may also involve a discussion of shallow versus deep penetration during sex. This is because some theories on sex selection believe that it is how close you deposit the sperm to the cervix that influences whether an X-bearing sperm or a Y-bearing sperm will be the first to greet the egg and fertilize it.

Therefore, if the Y-bearing sperm is faster, you'd choose positions to put it closest to the cervix, like rear-entry positions. The X-bearing sperm is slower but lives longer, so you can use positions like the missionary position to allow for a more shallow penetration. This supposedly works because the Y-bearing sperm won't be as viable by the time they swim that far to reach the egg for fertilization.

One mother, who wished not to be identified, says that she liked this because she didn't even have to tell her husband what she was doing. She just let him think she was being adventurous in bed. "No one was complaining, and I simply kept my secret to myself," she says with a smile.

The cost of this type of method is relatively low. You can look up information on the Internet for free, or make a one-time purchase of some reading materials. This allows you to use the method as many times as you want. The problem is that there is not really any science to back it up. This means that your chances are more likely close to the 50/50 chance you would have by doing nothing different. So in the end the cost may be very high financially if you wind up having more children than you may have originally planned in an attempt to influence the sex of your baby, and if you chose to keep going until it works. Or the costs can be very high emotionally if you stop having children before you achieve your goal of having a baby of a specific sex.

Diet and Supplements

While sound dietary and healthy habits surrounding nutrition are generally beneficial in getting pregnant, they have never been proven to affect the sex of the child being conceived. Still, this does not stop people from trying to tell you that you can influence the sex of your baby before conception by altering your diet (usually radically so).

There are those who would propose certain diets high in some foods and low in others. These so-called "preconception gender diets" are designed to change the pH balance of a woman's body, which manages a delicate balance between acidic and alkaline. These are considered fairly long-term methods, since you'd need to follow the specific diet for at least six weeks before attempting a pregnancy. Any less than that and you lose the potential effects on your body's pH level.

The theory about changing your body's pH suggests that the polarity of the human egg is affected by the change. But there is also a theory that it changes the environment in the vagina, making it more favorable to certain sperm—either X bearing or Y bearing, depending on the diet you chose. Other methods of directly attempting to change the vaginal pH are bandied about as well. The most typical method is douching.

While scientists have doubts about the successes of the diet method of sex selection, on one thing they agree—poor nutrition is a concern. There is always a risk that the mother-to-be will radically alter her diet to such a great extent that she either renders herself infertile or creates a very unhealthy environment for a pregnancy to exist.

Diets for Girls

According to various sources, if you would like to influence the sex of your baby in favor of having a girl, you will want to eat a diet high in calcium and magnesium. Some of the key points for this particular diet include an adequate intake of water, supposedly to help you absorb the calcium.

You will want to increase your intake of calcium-rich foods, like milk, cheese, yogurt, and so on, to increase your chances of conceiving a baby girl. You are also allowed to eat all forms of pasta, and brown or white rice. You should have no more than one egg a day, but one is acceptable. Meat should be limited to 4 to 5 ounces per day.

It is interesting to note that in the preconception diet for females, it is recommended that you avoid alcohol and caffeine. Whether this diet works for sex selection is yet to be seen, though this portion of the recommendation is perfect as a preconception diet to help prevent birth defects and increase natural fertility.

Diets for Boys

To push your chances of a baby boy, apparently women need a diet that's high in both sodium and potassium. The secondary goal is to lower the amount of calcium and magnesium in your diet, because these favor females.

Since dairy products are basically a no-go on the boy sex selection diet, you'll need to switch to alternative forms of dairy products. One example given is to use soy milk in cooking when it calls for regular milk. You should also avoid the yolks of eggs, but the whites are allowed, sparingly.

Meat may be unrestricted in the boy diet, but while on this diet you should try to keep your choices health conscious, like choosing grilled portions. You can also choose meats high in potassium and sodium. Kosher meats may be a great alternative to adding that extra salt in your diet, because of how they are cured in the koshering process. Liver is also a good option, if you can stand it.

Supplements for Sex Selection

Supplements are also a flavor-of-the-month type of sex selection method. Lucky is the vitamin or supplement manufacturer that gets the rumor started that its product helps with conception attempts for one sex or the other. Then seemingly obscure supplements will start flying off the shelf faster than the FDA can track the rumors on the Internet. The best recommendation, despite all the hype, is to stick with a balanced prenatal vitamin.

Supplements for a Girl

Like the diet prescription for increasing your chances of having a baby girl, the supplements theory encourages would-be moms to increase their calcium and magnesium. It suggests that you supplement with 800 mg of calcium, 300 mg of magnesium, and vitamin D.

Supplements for a Boy

Supplements can also be taken as a way to increase the desired potassium and sodium levels. It is recommended that you take about 200 mg of potassium every day. While it is true that certain vitamins can help a man, particularly if he is suffering from issues related to his sperm quality or counts, in no way are they associated with the sex of the baby that is conceived.

Vaginal pH

The altering of the vaginal pH level has been around for years as a theory in sex selection. Many popular notions of sex selection have tinkered with this idea at one point or another. Some people have recommended douches of vinegar or baking soda, depending on the sex preference.

Luckily, with the advent of the knowledge that douching actually does incredible harm to the normal pH of the vagina, many people have stopped recommending this form of altering the vaginal pH. Still, some die-hards out there continue advocating the use of certain douches to help shift the balance of your baby's sex to one side or the other.

The only difference in these newer douching instructions is the content. For example, the lime douche for girls is widely touted on some Internet message boards. The pH of lime juice is said to be between 2 and 3. In fact, in looking for more information, I also learned that lime juice as a douche is used by some as a method of birth control when combined with aloe vera gel, and it is also thought by some to protect against HIV infection.

Studies done on sex selection using agents to change the vaginal pH suggest that few of them work. Occasionally a statistic can be found that may be individually significant, but overall there was little change in the sex selection rates.

0+12

This method is only useful for conceiving a baby girl; it is not supposed to be used for conceiving a son. This method—called O+12 (pronounced "oh plus twelve"), where O stands for *ovulation* and not *orgasm*—was discovered by accident. It was only after a mother, identified only as "Kynzi," used another timing method, the Shettles Method, and conceived her sixth son, that she decided that the infamous Dr. Shettles had it all wrong.

Kynzi was reading a study from New Zealand that had been done to research the Shettles Method. This study had turned up some basic evidence that if you conceived 12 hours after ovulation, the results were more often girls than boys. This went against what Shettles said in his books. Some data also suggested that by abstaining from ejaculation for a period of time, more daughters were conceived. The theory is that there was a higher concentration of female sperm following this build-up of ejaculate.

The key to the O+12 method is to pinpoint ovulation. You do this by using various methods of ovulation prediction, including ovulation prediction kits (OPK), basal body temperature (BBT), and cervical mucous. You can also use other body signals like *mittelschmertz* (middle pain). This enables you to have sex after you ovulate.

The other part of the O+12 plan is the abstinence from ejaculation. This means no orgasm for your partner for up to a week before your suspected ovulation. This can be very difficult for some men, and it is one factor over which women have no control. Refraining from ejaculation is an attempt to build up the theorized increase in the population of X-bearing sperm.

Once you believe you have ovulated, going on the data you collected yourself, you will have sex to ejaculation one time only. This intercourse needs to be timed between 8 to 20 hours from ovulation to hit that "girl window." Then you are not to have sex again until you are positive you are no longer fertile. This is unacceptable to some couples, who choose to have sex after this time but instead of timing use barrier methods of birth control, like condoms.

It is hard to determine how effective this method is, though Kynzi, the woman who advocates its benefits, says that it's about 90 percent effective at giving you a baby girl. Considering that the method is one that you can use with "boosters," such as diet, douching, and others, it would be difficult to study its efficacy. Online surveys filled out by those trying the O+12 method provide varying results. None of the surveys comes anywhere near hitting the 90 percent success mark claimed by Kynzi, and some are even much higher for the boys than for girls.

Most basic texts on getting pregnant, regardless of which sex you're trying for, advise that once you've ovulated, your chances of getting pregnant go way down. While you have a lower overall fertility rate, there are a number of families who claim that by using the O+12 method you are virtually promised the baby girl of your dreams.

Lunar Theories

Theories abound concerning the moon and lunar calendars and their effect on sex selection. One version has it that if you conceive at certain points, the signs of the zodiac will indicate the sex of your baby. Therefore, you would simply look up which sex you wanted and have sexual intercourse to conceive under that sign.

According to this theory, Capricorn, Pisces, Taurus, Cancer, Virgo, and Scorpio are girl signs. The boy signs are Aquarius, Aries, Gemini, Leo, Libra, and Sagittarius.

Another theory of sex selection related to the moon is based on the lunar calendar. So if you ovulate during a new moon you would have a girl, whereas if you ovulate during a full moon you would have a boy.

Some studies out of Europe indicate that in fact the moon may have an effect on the pH of the endometrium, the lining of the uterus. This harks back to the theories that pH helps to influence the sex of a baby.

Questions to Ask about Low-Tech Sex Selection

If you are interested in using a low-tech sex selection method, you must gather information first. Here are some questions to ask of your proposed professional that will get you started on the information-gathering phase. You will need to modify some of these questions, based on which method you are considering, whom you are talking to, and other factors.

- What is involved with this method?
- If I need to buy anything, how much will it cost?
- Will I have to buy it more than once for each attempt or cycle?
- Will I have to buy it for subsequent attempts or cycles?
- Can I talk to anyone who has done it?
- Are there any formal complaints about your company, business, professional person, or method?
- What are the benefits of this method…
  —to me?
  —to my partner?
  —to my family?
- What are the risks of this method…
  —to me?
  —to my partner?
  —to my family?
- What are the alternatives to this method?

- If I ask for support in using this method…
  - —will the support come from professionals?
  - —will the support come from others who are trying it, or have tried it?
- Where do I file a record of failure or success?
- If I need to request a refund for an unsuccessful result that you have guaranteed…
  - —how do I do so?
  - —how many couples have had to do so?
  - —have you ever refused a refund, and if so, why?

5

The Shettles Method

Dr. Landrum Shettles, a New York gynecologist, was the father of many things involving fertility, as well as sex selection. After receiving his medical degree in 1943, he went about pursuing human reproductive issues. In 1950, he wrote *Ovum Humanum,* showing the human egg at various stages of development, something never seen before. He was completely devoted to his work and research, even attempting in vitro fertilization in 1973, before his superiors at Columbia Presbyterian Hospital stopped him and eventually fired him. After a series of jobs working in the obstetrical field, he eventually wound up in Las Vegas. He even briefly tried to start a human cloning lab. His pioneering research and dedication led to many of the great techniques we enjoy today in the field of fertility treatments. To that end, his sex selection method is one of the most well known and frequently discussed, even today, after his death in 2003 at the age of 93.

The method that Shettles proposed involves getting to know the menstrual cycle to help pinpoint ovulation. Once ovulation is pinpointed you will be instructed when to have intercourse, based on the factors that have been outlined earlier about the natural characteristics of sperm and the menstrual cycle.

Other timing methods are also used today, but that of Shettles is definitely more researched and used than any other. If you are going to look at using a timing method, you'll find a larger body of knowledge that goes along with the work of Landrum Shettles.

Shettles Basics

The Shettles Method revolves around the timing of intercourse to coincide with "girl days" or "boy days." In this theory, since the X-bearing, or girl, sperm can survive in the preovulatory conditions of the reproductive tract longer and tend to swim a bit slower, you can have sex within a broader window before ovulation to produce a girl.

If you are trying for a boy, you would do the opposite. This is because, according to Shettles, you are more likely to have a baby boy if you have sex at ovulation. This is because the Y-bearing, or boy, sperm can travel more quickly, but is less able to survive in the reproductive tract for long periods. So, for Shettles, the timing of both sex and ovulation is critical.

To time intercourse correctly, you use the cervical mucous method and basal body temperature (BBT) to predict ovulation.

In previous editions of the Shettles Method book, the authors advocated douching to help change the pH balance of the reproductive tract, but they have since backed off on this portion of the method, perhaps due to the knowledge that it is potentially harmful to the woman.

April always knew she wanted boys. It was her turbulent relationship with her mother that encouraged her not to want to repeat history. So when she was expecting her first baby and heard the words "It's a girl!" at her ultrasound, she felt shaken. What on earth was she going to do with a daughter?

While she had never considered sex selection, she had always just assumed that she would be the mother of boys. After getting over the shock, she figured this was her chance to change history and be the outstanding mother of a daughter.

> "I began having a really strong 'pull' toward having another girl. It would be so great for our first daughter to have a sister, and they could share clothes and toys and friends. So I approached the subject of purposefully trying for a girl

using the Shettles philosophy I had read about in many of my infertility books with my husband, and he was excited!"

They planned to watch for the day of ovulation and follow the Shettles Method to have a baby girl. To do this, they would monitor the cervical mucous to help pinpoint ovulation. They timed their intercourse, all the while avoiding "boy days" so that they could maximize their chances of another girl.

"We wanted to have our next child pretty much right away. That was our plan. Then secondary infertility was diagnosed for us, with no explanation. This time, though, we got a referral to a reproductive endocrinologist. He decided to do a round of Provera (when it was determined that I didn't ovulate the previous cycle), followed by one round of Clomid. With the information the doctor gave us and the research we did, we came to realize that with this course, we should know the very day I ovulate.

A few weeks later, at our follow-up appointment, we got the news that the Clomid worked and we were expecting another baby. We had one ultrasound done at week 18, and when this technician announced 'It's a girl!' I said, 'I know.'"

Whether it was the Shettles Method in this case, or the use of Clomid, which is also known to increase the odds of having a baby girl, we will never know. The good news is that April had the daughter she wanted with that cycle. Their next attempt produced their third child, this time a son.

Shettles for a Girl

The Shettles Method has been used time and time again. Some call it a poor man's MicroSort®, simply because of the low costs. In addition to minimal costs, however, there are also highly questionable outcomes.

The basics of using the Shettles Method to conceive a baby girl involve many steps. In fact, some of the literature available calls into question whether the number of steps involved in the Shettles Method negates any potential success it may have produced. Because the user of the method must follow *all* the steps, the number and level of difficulty of the steps may impede the ability of the end user to produce the desired results.

One of the numerous steps you must follow precisely in the Shettles Method is timing your intercourse. This means that you must meticulously track your cycles. A simple error on your part means a greatly lowered chance of having the baby of the desired sex.

You will want to learn how to pinpoint ovulation in your cycles. Shettles recommends that you do several practice cycles to figure out best when you ovulate. The problem here is that even by doing practice cycles you have no way of knowing whether you're actually pinpointing ovulation or are just practicing incorrectly every month. Those who have done Shettles recommend using your ovulation as the day that also matches your peak cervical mucous, so they say not to have sex on that day, but rather to mark it as your date of ovulation in your practice cycles.

Shettles said that by waiting until days before ovulation to have sex, you give the X-bearing sperm an advantage. This is because those sperm supposedly are a bit slower at swimming but are heartier survivors in the reproductive tract. He also advised not using hormonal forms of birth control during the practice cycles, because they can mask your body's natural fertility signs. Condoms are, however, generally acceptable as birth control during the practice cycles.

Once you have practiced for several cycles and feel confident that you can pinpoint your ovulation, you can begin in earnest. For a baby girl, you will want to have sex between 2 and 4 days before ovulation.

You must be very careful, though. If you have sex too soon, you won't get pregnant. Have it too close to ovulation, and you have switched the odds in favor of a boy.

In fact, when trying for a baby girl you can have sex frequently from the last day of your period. Shettles said that frequent ejaculation can lower the sperm count, which he claimed to favor girls. By "frequent" he meant up until your cut-off day.

This means that you need to establish a cut-off date—the day on which you will stop having sex during the rest of that given cycle, until several days after ovulation, when you are no longer fertile. On many sex selection discussion boards the discussion of cut-off dates gets quite heated, possibly because many women find it difficult to even set such a date—and that's even without polling the husbands!

For the actual sessions of intercourse, according to the Shettles Method special techniques can be used. One of these is related to the position during sexual intercourse. Shettles' claims that the missionary position is best to favor a baby girl are more about shallow penetration, as opposed to old wives' tales that talked about men choosing positions of domination. Shettles thought that the man-on-top position would help the X-bearing sperm win the race, because they are heartier and would last the extra-long journey to the egg.

This position, however, can be quite uncomfortable for both partners. It can also disrupt the rhythm of sex enough to make ejaculation impossible. Says C. K., the mother of several sons:

> "My husband had trouble on numerous occasions just trying to remember how far to go in and what to do when. We never got pregnant using Shettles simply because it drove my husband crazy and rendered him impotent."

Next on Shettles's list of things to avoid during sex when trying for a girl was the female orgasm. His claim was that the cervical fluid becomes more alkaline and plentiful when women have an orgasm. This means that her orgasm would at that point give preferentiality to the Y-bearing sperm, making it more likely that she would conceive a baby boy. This means that women should avoid orgasm during foreplay, intercourse, and even afterward. Kevin, a father of several daughters, reports:

> "I have problems having an orgasm if she can't—now that would really cause fertility problems. I want to pleasure my wife. There is just something fundamentally wrong with not being allowed to do that."

Kevin believes Shettles is wrong, saying:

> "The whole point of conceiving a child together is to have a beautiful process, and a part of that process is the orgasm. I hated the whole scripted thing. It was like it was out of some torture manual for husbands from yesteryear."

The vaginal environment has more than just the avoidance of orgasm to do with the conception of a girl. In fact, Shettles recommends that you make the environment as acidic as possible. Originally this method recommended douching with a premade vinegar douche about 15 minutes before sex. Now, however, citing the risks of vaginal douching, which include increased vaginal infections, potential ectopic pregnancy, increased risk of pelvic inflammatory disease, and other negative consequences, the Shettles book says to check with your doctor.

Obviously, other methods recommend changing the alkalinity of the vagina by diet and supplements. Shettles made no such dietary recommendation. But he did advise potential mothers wanting a baby girl to avoid caffeine, which he said is bad for a woman's fertility levels in general.

Shettles for a Boy

Having a baby boy with the Shettles Method is also a lengthy process. Once again, the doctor recommended that you learn to pinpoint ovulation—the core of his method for sex selection. With a boy, though, it is more about pinpointing than it is with a girl, where it is more avoidance. This means you have to be extra-accurate in knowing when you ovulate.

In preparation for your attempt to conceive a boy, you would have several factors to try to achieve. One was to have the father wear boxers, or in general keep his scrotal temperature a bit cooler. This also means avoiding hot tubs, not riding his bike, and the like, around the time of the desired conception. Although such practices don't only affect the Y-bearing sperm, they need all the help they can get to reach the egg.

The goal is to have one session of well-timed intercourse. Ideally this would be about 12 hours before the woman ovulated. The practice sessions for ovulation prediction truly come in handy on this one. For example, if you are using a charting method with basal body temperatures, you can learn to predict the temperature dip in your chart. This dip is when you would want to have sex. Some women also recommend using your cervical mucous as a back-up indicator. The recommendation is to have sex around the shift from the peak mucous to the thicker mucous, which helps pinpoint ovulation as precisely as possible.

You want to get as close as you can to ovulation. Remember, the egg has a very short life. Sperm have longer lives than the egg, though the Y-bearing sperm have shorter lives than the X-bearing ones. So if you have sex around the time of ovulation, the faster, Y-bearing sperm are waiting for the egg and are more likely to be the winning sperm when it comes to fertilization.

During your sexual attempt for a boy, the woman should have an orgasm. (Thank you, Dr. Shettles!) It is believed that since there is an increase in vaginal mucous after a woman climaxes, this favors conception in general, but also favors the faster, Y-bearing sperm. The woman's orgasm ideally should happen just before the man's, though if it occurs during foreplay that should work too.

Shettles also said that the man should build up his sperm, by not ejaculating for four to five days before trying to conceive a baby boy. The goal of this is to build up the sperm count. Shettles associated a higher sperm count with producing more boys, even if it did not change the actual number of Y-bearing sperm in the ejaculate sample. He recommended doing everything you can to achieve a healthy sperm count, such as regular exercise, proper nutrition, and avoiding illness and stress—all of which can reduce the sperm count.

Shettles hypothesized that the Y-bearing sperm had better odds in this scenario. The faster, Y-bearing sperm had a better chance of winning the fertilization race. Remember that in addition to being faster than the girl, or X-bearing, sperm, the Y-bearing sperm also die faster.

Thankfully, his recommendation to douche with baking soda was deleted from the latest edition of his books. He also did not advocate any type of special diet for either the man or the woman. But he did suggest that the man ingest some caffeine 15 to 30 minutes before sex to add a boost to the male sperm.

Why Shettles Isn't Your Best Bet

With all its complicated steps and complex formulas for douching, on first reading about it one might think that the Shettles Method was a high-tech method with a near-perfect record of sex selection. The truth, however, is that while Shettles is one of the oldest methods known to consumers, it's far from perfect and not remotely high-tech.

Numerous studies have both disproved and proven the Shettles Method of sex selection. Shettles himself vehemently supported his own theories and was known for openly attacking those who opposed him. These attacks included ones on Elizabeth Whelan, who wrote a book on sex selection that basically contradicted everything Dr. Shettles had said.

One study reported that even if the Shettles Method sometimes worked, the average couple wouldn't be able to get pregnant. This was partly because of the poor timing, particularly for baby girl attempts, and partly because the authors of the study believed that the steps were too difficult for most people to follow. This view can be noted time and time again on message boards and in talking to women who have used the Shettles Method. They often discuss how they "did everything except..." and then they fill in the blank. The lack of female orgasm can be a big problem for some people. Even after all the contradictory reviews and experiences, the trials of following the Shettles Method to a T are done by many—though often without success, despite the self-reporting from Shettles and his followers.

Diane shares her story:

> "Part of our discussion when considering becoming parents was that there were already six granddaughters and no grandsons on my husband's side of the family. I felt the odds might be against us, since in my family there had always been a girl born first."

This myth still abounds, though it has now been shown that having only babies of one sex in the larger family group doesn't raise the rates significantly of your having a baby of that same sex. Diane recalls:

> "In my search for knowledge on conception, I came across a book called Your Baby's Sex: Now You Can Choose, by Landrum Shettles and David Rorvik. It sounded like just the thing to increase our odds of a baby boy."

There are couples who report feeling anxious about their ability to conceive. Since using Shettles can lower the possibility of conception, this fear can be well founded in some couples.

It was something that Diane and her husband thought about too.

> "We waited and waited and got a little anxious about the prospects of me getting pregnant. Along came April 1, 1980. My husband had worked long hours the

> day before so when he arrived home he had been without sleep for about 35 hours. I meet him at the door with a silly grin and a 'Welcome home, sailor! Today's the day.' Even though he was exhausted he marched into the shower and I marched into the other bathroom to douche with a baking soda solution and we met in the bedroom."

Unable to take advantage of ultrasound technology to find out whether her Shettles attempt for a baby boy had worked, Diane had to wait for the birth. After a long, exhausting labor she had this to say: "Thank goodness I have a strong husband, since the next six hours were long and difficult. Finally at 6:38 p.m. after pushing for several hours and getting third-degree episiotomies my baby came into the world—all 10.5 pounds and 24 inches of her.

> "After all the charting, the holding off on intercourse, the douching, and the special positions, our baby was a beautiful little girl. So did the Shettles Method work for us? Nope. Allison was granddaughter number seven for my husband's family and the first grandchild for my family."

Diane completed her family with a son two years later, with no further attempts at sex selection. She is quick to point out that he is "the only boy from our generation of nine nieces."

Nevertheless, the Shettles Method is very popular. Even though Shettles is much less expensive than MicroSort®, you must consider the highly questionable outcomes when choosing your sex selection method.

Is Sex Selection Right for Your Family?

Make Sure You Want a Baby

Choosing to try sex selection makes it far too easy for you to get caught up in the hype. You can quickly be swallowed up by the obsessions of others, even if they were not originally your own.

The topic spurs many a debate, because more often than not it offers the potential answer to a longing.

One of the things said by many parents who have gone for sex selection was that they had to make sure they truly wanted a baby. The biggest fear most parents reported was how they would feel, should they get the "opposite" of what they were looking for in a baby. This sentiment was echoed even more by those who had received the blessing of an opposite. JoJo, now pregnant with her third son, remembers:

> "Using the Ovacue® with the Shettles Method had me excited to try for my girl. I timed it with a counselor at Zetek (the company that makes the Ovacue®) and we decided that I would have sex every day until three days or so before ovulation. Well, I got pregnant the first month I tried, and I ran into some interesting obstacles. Even though the Ovacue® told me I would ovulate on June 4th, my body decided to ovulate earlier on June 2nd. This only gave me about a two and a half day cut-off, and I'm sure boys [boy sperm] were still alive to fertilize the egg. I never expected to get pregnant so quickly."

Originally she thought she should have done some things differently:

> "I regret not using supplements to sway the odds even more. If I had, I don't think I would have needed a forum for support. I'd have a girl on the way. Of course nothing is foolproof. I was anxious to know, so at eight weeks I contacted the Pink or Blue company and ordered my early gender test. They had a problem reading my sample and had two inconclusive results before telling me on the third try that it was indeed a boy."

JoJo and her husband have decided that the next time they have a baby, they will use a high-tech method of sex selection. She says that her husband prefers to use the MicroSort® method because of its accuracy.

Dr. Mark Perloe, a reproductive endocrinologist who does sex selection, voices his thoughts:

"I get concerned about family balancing and the issues surrounding how the couple will respond if they seek family balancing and end up with a child of the 'wrong' sex. Parenting is about learning to adjust to the hand you are dealt. I think whenever we place our personal expectations on our offspring, we will ultimately be disappointed. Parenting is about discovering that these young people create themselves and the best we can do is guide them to become the people we hope they will become. I am just a bit concerned that some couples seeking sex selection have expectations that will interfere with their being accepting supportive parents."

Sally, a woman who wants another baby, recommends:

"Make sure that you first and foremost want a baby. I remember learning this, thankfully through another couple. They had two daughters and really wanted a baby boy. They did nothing to influence, not even one of the goofy methods. They had twin daughters. The woman was beside herself. It made it really hard for her to attach to them. I really want another baby, and I would just like to sway the odds in favor of a girl."

Most couples who have babies that are opposites of the original plan can eventually find peace with the results. Some will choose to take a high-tech route in the future, while others will just accept the family that they have and choose to be done with future children. Most of these parents say they wish they had thought about this part of the process earlier. Amanda admits:

"I gave it a cursory glance. Thankfully, when my son was born my initial thoughts were good ones, even though I was expecting a daughter. It didn't affect me too much during his babyhood, though when I finally did have my daughter it did make me wonder a bit more."

Need Versus Want

Plenty of couples choose sex selection because they would like to have a balanced family. They want to have the experience of raising daughters and sons. This is a case of "want."

Sex-linked genetic diseases make the goal of sex selection a "need" rather than a want for some couples. This dictates that they will have to risk having a child that is afflicted by a terrible disease, or even a carrier of such a disease, if they wish to become parents. The beauty of fertility science is that this issue has been resolved. Would-be parents are now able to have the child of their dreams and be free from worry about its health or the health of future grandchildren.

Sex selection for these couples is a high-tech issue. They do not have the time or luxury of wasting the chance of ensuring their baby's health on risky behaviors like the low-tech sex selection methods available. They need to actually be able to accurately choose a baby who is healthy. These couples do not have the luxury of family balancing without the use of high-tech methods such as preimplantation genetic diagnosis (PGD).

For these couples, there is little controversy about the use of sex selection. Most physicians and individuals would give their complete blessing to spare such families and children the horrors of disease. Dr. Perloe adds:

> "We do not participate in family balancing by PGD but will support or help facilitate use of MicroSort® combined with IVF for family balancing. We feel more comfortable using MicroSort® combined with PGD for medical conditions."

Low-Tech Failures

Since low-tech methods of sex selection like Shettles, positioning, diets, and supplements are much less likely to work in your favor, they are therefore also much more likely to cause a pregnancy to be of the opposite sex. Couples choosing to use these methods must know this, when they start undergoing the process.

By not recognizing the risks that your family is taking by using these low-tech methods, you overlook the risks of the impact on your family of having a baby of the opposite sex. If you enter into the low-tech methods knowing what you're up against, then the risks of your having an emotional upset are lowered.

Not only can the use of the inaccurate, low-tech methods of sex selection cause you emotional turmoil, but also it can wreak havoc on the rest of your life. Many families go on to use high-tech methods of sex selection in subsequent pregnancies. For these attempts they spend money that could have been directed more wisely the first time. One mother who wishes to remain unidentified says:

> "I love all of my children. But we definitely hadn't planned on three. I wish we had simply paid the money for the doctors to help us the first time. Financially, it would have been easier all the way around."

These low-tech methods should only be used for nonserious sex selection attempts. The goal should be to have a baby with some fun attempts to sway the odds one way over another. It's important that families who are very serious about the goal of selecting their baby's sex use methods proven to shift the odds firmly in their favor, like the Ericsson Albumin Method, MicroSort®, and preimplantation genetic diagnosis (PGD).

Put Your Money Where Your Uterus Is

While by definition sex selection may be a costly venture, there's always someone out there trying to make more money with it. Whether in the virtual world of the Internet or in brick-and-mortar offices in real life, scams abound. There are people who make a living preying on women who put their heart and soul into something as important as having a child.

When you have families who are so totally involved in something like sex selection, you have vulnerable people. They are often easily swayed and tempted by schemes to "get what you want quickly and easily." These schemes are often hurtful, both mentally and emotionally, to these families.

One of the best examples is actually on the fringe of the sex selection community. A company claimed to make a product that would tell you the sex of your baby at just a few weeks past conception. This sounds great, particularly if you've been trying to select the sex of your baby. Its test costs a lot of money. But women want to know that their hard work and efforts have paid off, so that they can either cope or celebrate. The test revealed answers that were the opposite of what the ultrasounds (done by the women's regular physician) revealed. When the women asked to have their money refunded, not to mention some relief from the emotional turmoil they lived through, the response from the staff was that the women were carrying babies with genetic defects, which threw off the results of the test.

This is but one example of money being taken from many women in exchange for false hope. It's no coincidence that most women have run into something like this at one time or another. Families in such emotional need are easy prey. Protect yourself.

High Tech Is Accurate, but Costly

The good news for families who need or want to have a baby of a specific sex is that it is possible. Not only is it possible, it is probable that they will succeed in having the baby of their dreams. As with all dreams, though, there is a price tag attached.

As you might expect, you will pay for the accuracy. The more accurate a method of sex selection, the more it will cost you. Whether you're spending a few hundred dollars for something like the Ericsson Albumin Method or several thousand dollars for the combined techniques of MicroSort® and preimplantation genetic diagnosis, most families will agree—where there is a will, there is a way.

Many mothers and fathers talk about the creative financing they go through to make these dreams come true. A number of them resort to alternatives of some sort to finance the pregnancy attempts. They may make more than one attempt; in fact, many women will require multiple attempts to get pregnant. This is true even for women who are naturally very fertile and have no trouble conceiving.

Boys Versus Girls

The perception is commonly held that if you could choose the sex of your baby, you would choose a boy. This comes from a historical perspective in which male children were often preferred. In certain communities and cultures, this is still the case.

Some countries even have laws about how many children you can bring into the world. The financial and emotional costs of this are enormous. To then have a female is looked down on. This is obvious, especially since some countries have had to outlaw not only sex selection, but even the ability to learn the sex of your baby before birth.

These countries have actually seen a shift in the balance of the sexes. The boys now outnumber girls by a good measure. This has led to problems in those countries.

The problem here is not the fact that people have choice, but that they have no choice. Societal standards vary widely. Some countries value females more than males, and vice versa.

Here in the United States there are a wide variety of couples making vastly different choices. While one might assume that boys are all the rage, they would be missing the buzz on the Internet and beyond—that girls are now the trend.

Currently many families are desiring baby girls. This winds up being very good for these parents because some of the high-tech sex selection techniques are extremely accurate for producing baby girls (though baby boys have their say too!). Says Dr. Perloe:

> "Certain populations come in for male offspring for the first baby. That still occurs, but as more of the sex-linked conditions are a factor in men, medical sex selection would make a female more appropriate."

6

High- or Low-Tech: Which Is Right for You?

There's no shortage of low-tech sex selection methods available. All you have to do is get on the Internet and search for the term "sex selection" to come up with tens of millions of pages containing theories on how to influence the sex of your future children. Many women like Amanda find it overwhelming to even ponder where to begin:

> "I just always assumed that there was one answer per sex. I figured a girl had a certain thing you did and a boy had another. I flipped out when I saw everything. I simply didn't know where to start or whom to believe."

This is often the case with people who are just beginning their search for information on sex selection. With all the methods out there, each one claiming to be more credible than the rest, you have a rocky road ahead of you. But we can offer some handy pointers to the real deal in matters of sex selection.

How to Tell If the Program Is Real

Don't get fooled by customer testimonials. There is nothing like reading letter after letter from supposedly real, satisfied customers to make you feel all warm and fuzzy inside. This is particularly true when these letters are accompanied by cute pictures of smiling babies and great stories of adversities overcome by whatever method you are currently considering.

The truth is that these letters may be great, but in fact they are fluff. They are basically anecdotal evidence. You should always ask to see the scientific data on the sex selection method being offered. Then ask whether there are any studies that refute their findings. Remember, though, that not everything that looks like science is good science. When in doubt, take the study to a physician, midwife, or scientist to help you interpret what you are looking at. Look for peer-reviewed information.

If it sounds too good to be true, it probably is. "A 200 percent money-back guarantee that our product will help you have the baby of your dreams!" If claims like these sound too good to be true, if they say you have nothing to lose, and if they make promises that sound really hard to keep, then they are probably not going to be helpful.

When a business makes a wild claim, like a 200 percent promise to return your money if you're not satisfied, or some such claim, my first instinct is to call them on it. There are ways to find out if anyone has ever used its guarantee. I usually start with the Better Business Bureau for the community or region. You can look up businesses confidentially online to get some basic information. You can also search online for customers, both unhappy and happy ones. Blogs are really great platforms to find honest opinions so that you can compile information and make an informed decision.

Be leery about paying large sums of money over the Internet to companies promising goods. Be sure that you have done your research before you send any money to any company over the Internet. This is true of small amounts too, but it is always more problematic for larger sums of money. Ask yourself questions about what you are expecting for your money. Do you get anything tangible? Does the company have a physical presence in the United States? Is there a money-back guarantee?

When in doubt, just don't do it. If you do spend money on an Internet site, be sure that you cover yourself in some manner. If you use certain forms of payment like PayPal or your credit card, you may have some protection.

There's no such thing as a free lunch. This is as true in sex selection as it is in life. The advice that you're given for free from others on the Internet is often well meaning, though not necessarily always useful. In fact, it may be dangerous. Be sure to check everything out with your personal physician.

Clinical trials are the gold standard. When you're checking out the efficacy of a method, or how well a theory holds up in practice, look for a clinical trial that supports it. This usually means that the study is very well thought out and monitored by external sources with no monetary gain or other investment in the company or the theory.

Some methods have already had clinical trials, or are currently in such trials. The latter should not concern you. The Institutional Review Board (IRB) monitors any studies done on humans. This external body is the governing agent for the study being conducted. It can provide you with information on the study or trial being conducted and serves as a repository for complaints lodged against the trial or study.

Look for programs that admit their weaknesses. There is something charming and believable about a program that can say that "such-and-such is our weak spot, but we're working on it." This might be a program that reports that it has lower stats for a certain sex than another. Or it may be a clinic that offers preimplantation genetic diagnosis while reminding you that such a thing is still considered experimental in nature.

Are You a Low-Tech Tragedy Waiting to Happen?

The low-tech methods of sex selection seem fairly enticing. For most of the methods out there, the price is right—from inexpensive to free. There are few, if any, embarrassing questions to answer or exams to endure. You don't even have to tell your husband or partner, if you choose not to do so.

They can sound like a dream come true. And they seem perfect, if your goal is to have another baby to love. If, however, you are really tied to the thought of having a baby of a specific sex, then you're very likely to be disappointed.

The good news is that the basic risk you take of having a baby of either sex in any cycle is about 50/50.

As you can tell from a brief review of many of these low-tech sex selection methods, a lot of them have theories that refute one another. The way to have a baby boy with one, it seems, is the same as to have a baby girl with another. The stories of their success are often told, but they are just that—stories. None of these techniques will offer you and your family the near-guaranteed rates that the high-tech solutions offer.

For these reasons, it is important that you consider which approach is the best option for you and your family. Is the low-tech, less-expensive, and less-reliable method the right option for you? Or would you be happier and financially more sound in the long run by choosing the high-tech methods that are more likely to prove effective, though at a greater initial cost?

How to Choose a Sex Selection Method That's Right for Your Family

Having a baby is supposed to be the happy aftermath of a quiet, mysterious, exciting coupling between a man and a woman. When the need or desire for a baby of a specific sex comes into play, that privacy goes out the window. That said, there is no single specific method that is perfect for every family.

Both partners must answer all the questions about their emotional, mental, and financial resources. Each of these resources will need to be cataloged for a couple to figure out if they're able to withstand whichever method they choose.

In particular, you may want to ask yourself the following questions:

Mental

- Do you understand the science behind the method of sex selection you have chosen?
- What is the potential success rate for getting the sex of the baby you want with each method?
- Are there any special concerns you have about your personal participation in each method?
- What's the pregnancy rate for the method you have chosen?
- What is the average number of cycles it takes to become pregnant with this method?
- Do you have any additional concerns or requests?
- Do you know the person to go to with science questions?
- What are the physical risks to you? Your partner? Your baby?
- Do you have a contact person at the clinic or with the method you have chosen?

Emotional

- How invested are you in the sex selection process?
- How tied are you to having the sex of your next baby be the one you choose?
- Do you have a plan in place if the baby's sex turns out to be not the one you have chosen?
- Will you share the information of your attempt with anyone?
- Will you find out the sex of the baby before birth?
- Are you planning to do genetic testing during pregnancy?

Financial

- How much money are you able to spend to do sex selection?
- Will spending this amount hurt your family financially?
- Will you have to go into debt?
- Do you qualify for any financial assistance because of genetic issues, or will you have any type of insurance support?
- Will you be able to pay the add-on costs, like the rush purity fees associated with the MicroSort® process, intracytoplasmic sperm injection, additional medical monitoring, and the like, if you desire or require them?

PART THREE

High-Tech Methods of Sex Selection

7

Using the Ericsson Albumin Method

History

Ronald Ericsson, Ph.D., is a pioneer in sex selection. Having worked with sex selection for over 40 years, he holds multiple patents. And having written many articles on topics relating to fertility, infertility, and sex selection, he is definitely knowledgeable in the field of sex selection. The good news for couples is that after four decades, Dr. Ericsson is still hard at work at his company, Gametrics Limited, helping to provide sex selection services.

Dr. Ericsson is passionate about making sex selection available to everyone, even those without other children. He says:

> "The decision to have or not have children is the right of parents. This is a human rights issue and others should not have the right to select certain criteria as to whether or not couples have the right to preselect the sex of a wanted child. To do otherwise would open up sex selection to individual and organizational biases."

One of the issues with many of the high-tech sex selection methods is that they are not available to everyone who wants to use them. There is always someone making decisions at some level about who can and who can't use the procedures. One benefit of using the Ericsson centers is their open policy on the use of sex selection. Dr. Ericsson's beliefs about sex selection remove those barriers for families. You can use the Ericsson Albumin Method for any reason, not just a few reasons that are deemed socially acceptable.

Some 18 centers in 9 states and 15 countries currently offer the Ericsson Albumin Method of sex selection. This makes it more widely available than some of the other high-tech sex selection methods. In the past, each of these centers purchased the rights to the process. Whether these licensed centers provide Dr. Ericsson with the results is unclear.

Dr. Ericsson says that many fertility centers are choosing to advertise that they use "Ericsson Methods" but are not actually a part of Gametrics. The patents have long expired and people are allowed to use similar technologies. The problem with this, according to Dr. Ericsson, is that they are not providing exactly the same service and protocol that his method provides; they're just using his name to legitimize themselves. Dr. Ericsson warns:

> "What they are offering is unknown to me as we do not train their clinics and I am certain the technicians are not using the same protocols we used as these protocols have changed with time and clinical experiences."

So if it's Ericsson's methods you want to use, *buyer beware.* Make sure you ask the center of your choice whether it has a specific business agreement with Gametrics. To ensure that its staff are in fact using the proper Ericsson protocols, you might also ask where they received their training for the methods they use.

The Process

Dr. Ericsson first described the Ericsson Albumin Method in the early 1970s. This consists of a multistep process of separating sperm through albumin, a protein found in blood that regulates the osmotic pressure. Over the years, the albumin method has had a good track record for helping to filter sperm to achieve a baby of the desired sex. The success rates for baby boys is between 78 and 85 percent.

Dr. Ericsson and Dr. Andrew Silverman have found that adding clomiphene citrate (Clomid®) for ovulation induction to the cycle can boost your odds of having a baby girl by up to 25 percent. This means that if it is a girl you are dreaming of, the success rates are 73 to 75 percent. For this reason, Clomid® is not used in cycles where the mother is trying to conceive a boy. Dr. Ericsson explains:

> "There are a number of papers in the scientific literature that provide evidence that the hormonal induction of ovulation will skew the sex ratio from the normal majority of males to a majority of females. The reason for this shift in the sex ratio remains unexplained. We use the drug clomiphene citrate as part of the protocol for female selection. It cannot be used for male selection as this would negate the increase for males. We learned through clinical trials that the use of clomiphene citrate in conjunction with sperm isolation will increase the percent of females by 25 percent. It should be noted that the MicroSort® method also uses hormonal induction of ovulation as part of their protocols for selecting either a male or a female. Part of the increase in females could be due to the use of the hormonal induction of ovulation. The low percentage increase for males achieved by MicroSort® may be due, in part, [to] the use of these drugs."

While these rates seem to be great, there are those who absolutely insist that they are just not real. Numerous studies have been done trying to prove or disprove the Ericsson theories. In fact, about once a year for several years, an article would appear that tried to debunk Ericsson and his method of sex selection. Many of these studies have shown that the percentages of sperm at the end of the process are not radically different from the 50/50 that nature provides. This is no surprise.

To be sure, the Ericsson Albumin Method does not claim that there will be a greater concentration of any one type of sperm—X bearing or Y bearing—at the end of the process. Ericsson does point out, however, that the dead and less than perfect sperm are lost in the process. This is true of sperm of both sexes, not just one or the other. According to Dr. Ericsson:

> "Our technology does not use a machine nor do we stain the sperm. The isolation process is based on the ability of sperm to swim with difference degrees of velocity. Human sperm that contain the X chromosome are known to be larger and have a longer tail. Therefore, we capitalize on these anatomical differences and place sperm [in] an environment that will magnify the different swimming ability of the X and Y sperm. This process is shorter and more sperm of higher quality are recovered for insemination."

Any couple interested in using this method of sex selection need only contact a fertility center of their choice that uses the Ericsson Albumin Method. They may choose a center on the basis of location, cost, or success rates, or a combination of the three. There is no one center that meets the needs of all couples seeking sex selection support. All the centers are independently operated and are not under the direct supervision of Ericsson himself.

Once the couple selects the center they want to use, there is usually a consultation. It can take place in person or by phone, depending on where you live in proximity to the center. The purpose is to make sure that the couple is healthy and willing to go through the process. A discussion is also had about the procedures involved, as well as the success rates. This includes a frank explanation of the failure rates as well. Naturally, the failure rate is not something that most couples want to think about.

The discussion will also include the length of time to pregnancy. The wait times to inseminations and cycles are typically not as long as with other methods. It can also take more than one cycle to achieve any pregnancy.

After a medical history and evaluation are done, the timing of your sex selection cycle is discussed. The choice of cycles will be explained. Instructions will be given on collecting the sperm sample for processing. Typically, a sample takes about three or four hours to be processed before the insemination can take place. This is usually done all in one day, since the insemination itself takes only a few minutes.

It is also important to point out that the Ericsson Albumin Method does not harm the pregnancy in any way. Its use carries no increased risk of miscarriage, nor does there appear to be any increased risk of the baby having a birth defect.

To begin the process of sex selection using the Ericsson Albumin Method, you need a sperm sample. There is nothing special about the gathering of this semen sample—it is obtained the same as any other semen sample, usually through masturbation. The special processing that this method employs will be done at the lab. The key is that the semen sample must be fresh. Frozen samples begin to deteriorate too much to be useful for the purposes of sex selection or to create a pregnancy.

Ericsson points out the key elements this way:

> "The important factors are: fresh semen of good quality and a known time for ovulation. The couple later informs the sperm center as to whether conception occurred or did not occur. If conception did occur, the couple is monitored to learn the details of the pregnancy and sex of the child."

The semen is washed and placed into a centrifuge, a machine that basically spins the sperm. Its force separates the seminal fluid from the sperm. From there the laboratory technician layers the centrifuged sperm, in a known quantity, over an albumin solution in a vertical column.

Ericsson proposes that the anatomic difference in viable versus nonviable sperm can be used to separate out the "best" sperm. The centrifuged sperm swim downward though the albumin solution. This race is not for the weak or the weary, and only the healthiest sperm will make it to the bottom of the column. The rest of the sperm are discarded.

The collected sperm are placed in a medium to prepare for artificial insemination or intrauterine insemination (IUI). The insemination is done near ovulation. Dr. Ericsson also advises the use of clomiphene citrate (Clomid®) for couples who are using sex selection to have a girl. Even in light of the fertility issues, this is something that he will *not* use when a couple is trying to sex select for a boy, because of the added possibility of having a girl.

The intrauterine insemination is done on the same day at the fertility clinic. The whole process normally takes just a couple of hours. Then the couple is out the door on the way home to await results.

Success Rates of the Ericsson Albumin Method

Once the insemination is over, the couple begins what is known as the two-week wait, the period of time between possible conception and either the start of your next menstrual cycle or the results of a positive pregnancy test. It is during this time that typically no tests or medical procedures are needed. For some couples, this can be a very long and frustrating phase, even when no sex selection process is involved.

As with any other form of fertility assistance, the success rates also depend on the pregnancy rates. It can take a woman several cycles to conceive using the Ericsson Albumin Method. This can be a result of the decreased number of sperm left after the technique is applied. Dr. Ericsson says that the remaining sperm are of the highest quality and points to the fact that human reproduction is itself imperfect, with only 20 percent of couples conceiving in any one cycle of naturally attempted conception.

As a result, the Ericsson Albumin Method may require multiple cycles to achieve success. It takes an average of 2.5 cycles to get pregnant if you have what is considered normal fertility and are age 40 or under. For couples in which the woman is over 40, the number goes up on average for the number of cycles it takes to conceive. As Dr. Ericsson puts it:

> "There is too much variability to state the number of insemination to bring about a pregnancy in women over the age of 40. Suffice it to say some conceive on the first insemination and some never conceive. The overall number of attempts to conceive does go up with increasing age."

With all this talk of sperm spinning, your head itself might be swimming! This can lead you to ask the honest question of how this method differs from other technologies. MicroSort® is the one most often compared to the Ericsson Method. Dr. Ericcson states:

> "The end product of an Ericsson Method is not an enriched sample. This means that there is not an increased number of X-bearing or Y-bearing sperm."

Dr. Ericsson is quick to point out what his detractors enjoy saying:

> "In the studies that do support the claims that this method increases the rates of male birth, they still have fairly even numbers of female to male ratio in the final sperm product."

While the description of the process sheds little light on how the process actually works, and while some scientists claim that it makes no sense at all, Dr. Ericsson has the statistics to show that his method does indeed work. This method of sex selection was the first high-tech sex selection method available and it remains on the market many years after its advent.

Costs of an Ericsson Albumin Cycle

Considering the sheer number of centers that employ the Ericsson Albumin Method of sex selection, it's difficult to give an exact cost. In general, the cost quoted usually includes the following services:

- Consultation

- Washing and processing of the semen sample
- Intrauterine insemination (IUI)

Things not covered in the cost will be extra medication like Clomid®, ultrasound tests, monitoring, pregnancy testing, and other extras. When trying to calculate what the total cost will be before you become pregnant, you need to look at any travel and travel-related expenses that are necessary and how many cycles you will probably need in order to become pregnant. The good news is that many centers offer this service both nationally and internationally, likely giving you a convenient choice of clinics, depending on your location and preferred options.

8

Using MicroSort®

History

MicroSort® is one of the more popular high-tech methods of sex selection. The biggest reason for its demand is mostly the amazing success rates that this sex selection procedure achieves. This method is also largely available to the general population, a fact that increases its popularity.

Since the sex of your child is determined by the sperm, that is where MicroSort® focuses. Clinicians using the MicroSort® process must be able to do three things to achieve the goal of helping you conceive a baby of the sex you want:

- Be able to tell the difference between the X-bearing sperm and the Y-bearing sperm
- Separate the two types of sperm
- Figure out which sample is X-bearing sperm and which is Y-bearing sperm

Dr. Lawrence Johnson originally developed a machine for use in animals with the U.S. Department of Agriculture (USDA), calling it a flow cytometer. The Genetics & IVF Institute (GIVF) worked with Dr. Johnson to modify this technology for use in humans to help families select the sex of their baby.

How MicroSort® Works

In general terms, the average semen sample consists of about 50 percent X-bearing sperm and 50 percent Y-bearing sperm.

Naturally then, your chances in any one cycle of conceiving a boy or a girl are about even. The purpose of the separation is to increase those chances in favor of one sex or the other.

We know that X-bearing sperm actually contain about 2.8 percent more DNA than the Y-bearing sperm, making them slightly larger. Since we know that the X-bearing sperm are larger, technicians using MicroSort® can separate the semen sample into X-bearing and Y-bearing sperm. The two types are sorted individually by use of the flow cytometer.

Flow cytometry is a process in which a machine called a flow cytometer is used to measure the difference in a group of cells. To do this, the cells must be tagged with a light-sensitive dye, a process called fluorescence in situ hybridization, or FISH. The FISH process is a type of DNA analysis.

Fluorescence in situ hybridization is done by attaching DNA probes to the X or Y chromosome. A Y-bearing sperm will have a green appearance, while the X-bearing sperm will appear to be pink or red when passed through a specific laser beam. The sperm are then counted and sorted according to these findings. Once this is done, the scientists and lab technicians can determine how much of the sperm in the sample are of the X-bearing or Y-bearing variety.

An enriched sperm sample is the end product of the flow cytometry process. This sample is concentrated with the X-bearing or Y-bearing sperm, depending on the request of the family choosing the sex selection technique.

The percentage of the sample that's of the desired sex is referred to as its "purity." That purity can vary from sort to sort. This is true even in the same sperm sample. This occurs because the range of what happens to the individual sperm is so great. A small portion of the sample is tested to verify its purity. No one really knows why some samples are purer than others.

Typically a report on the enriched sample's purity is provided to the couple a few days after the insemination or in vitro process is completed. You have the option of paying for a rush purity score, which can be made the same day as your sort. So if the numbers from your specific sort are not as high as you'd like in the favor of the sex you've selected, you have the option of discontinuing the cycle and trying again at another time. This is obviously a difficult decision to make.

Once you have decided to go through with the MicroSort® process, however, you will make your sex selection. The process through which sperm is sorted to increase your chances of having a girl is known as an XSORT®. A YSORT® would be the flip side of the coin, used to increase your odds of having a boy.

MicroSort® Success Rates

While MicroSort® has proven to be highly successful, it is nevertheless necessary to do medically sound studies to demonstrate not only its effectiveness but also its safety in humans. Clinical trials are required by the Food and Drug Administration (FDA) to give consumers confidence in the procedure and to protect them. Dr. David Karabinus, the MicroSort® Division scientific director at Genetics & IVF Institute, reports:

> "The clinical trial has approval to enroll 4,500 couples and for 1,050 babies. Currently, approximately 850 babies have been born. Our focus at this time is completing the trial so the data may be submitted to the FDA for their review and approval."

By doing the clinical trials of MicroSort®, the Genetics & IVF Institute hopes to establish the efficacy rates of MicroSort®. The first clinical trials were actually completed with swine and rabbits. Once safety in several generations of faster-breeding species had been proven, permission was granted for human trials. The Genetics

& IVF Institute and its sister site at the Huntington Reproductive Center in California are constantly being watched during the trial by the Institutional Review Board (IRB) to ensure that this approved trial is proceeding as expected while the MicroSort® technology is studied in humans.

The great thing for consumers is that this clinical trial gives would-be parents an extra boost in confidence. Let's face it, this is a highly personal, not-inexpensive venture to hopefully lead to a desired end—a child of a specific sex, either for prevention of sex-linked genetic diseases or for family balancing. Couples using the technique appreciate being able to see the results in black and white, even as they're constantly being updated. Enough data is provided (and it's not anecdotal data) to help them make a decision whether to go forward.

You can always check out the most recent data by calling Genetics & IVF Institute or by going to its website. The most current statistics are broken down by XSORT® and YSORT®.

The XSORT® has an average enriched sample that contains 88 percent X-bearing, or girl, sperm. Of the XSORT® babies, 91 percent were females at birth. This is out of 574 pregnancies, making it 525 baby girls.

The YSORT® samples have a 73 percent average for the boy, or Y-bearing, sperm. So 76 percent of the 152 pregnancies in an attempt to have a boy were actually male (127 boys born). *(Data accurate as of November 2006 per microsort.net.)*

These rates far surpass any sex selection method that can be done at home. Indeed, these are extremely high success rates for *any* sex selection technique, particularly when not using in vitro fertilization and preimplantation genetic diagnosis. This opens sex selection up to many more couples who might otherwise become frustrated by trying ineffective methods to have a baby of a specific sex.

The risks to the mother and baby are quite small when using this technology. Also, no medications are necessary if the couple chooses to do a natural intrauterine insemination cycle. All things considered, this is the most effective sex selection method, with the lowest price tag.

The MicroSort® technique has been hailed as amazing by many in the sex selection community. They praise to the skies its ability to offer sex selection to families who may not have been able to have a child of a specific sex without it.

How Is the MicroSort® Process Done?

MicroSort® works in a couple of different ways, each with its own benefits and suboptions. One of the most basic decisions you will need to make is whether you would like to do MicroSort® using the intrauterine insemination (IUI) method or with the use of in vitro fertilization (IVF).

An intrauterine insemination (IUI) is done after you figure out precisely when you ovulate. Then you will travel to either the Genetics & IVF Institute in Virginia or the Huntington Reproductive Center in California to be there for that time. Your husband or partner will leave a sperm sample for sorting. You'll return on the appointed day to be inseminated with the enriched sample for the chosen sex you have selected. Then you return home to wait for the results of your pregnancy test.

The least medically invasive option is the natural intrauterine insemination option. This option does not include medications to induce ovulation or to increase the number of eggs. It is normally the option chosen by women of normal fertility. Some would argue, though, that as long as you are going to do an insemination, you might as well use the ovulation induction methods.

Some risks are associated with using ovulation-inducing medications, like Clomid®. Physical side effects can include mood swings, headaches, enlarged ovaries, and other issues. The biggest physical risk is that of a multiple pregnancy, specifically twins. Carrying multiples puts the mother and the babies at a greater risk during the pregnancy, and beyond.

A medicated intrauterine insemination (IUI) cycle to help you ovulate may be recommended if you have fertility issues or problems with ovulation, though a goodly number of MicroSort® patients have chosen to use ovulation-induction medications despite having no previous fertility issues. This will be a decision that you can make in conjunction with your husband or partner and the reproductive endocrinologist.

You also have the ability to use donor sperm in your cycle. This would be an option for couples who have severe male factor infertility but no female infertility diagnosed. Some couples choose to use this route if the male partner has already had a vasectomy but they have now decided to have another baby.

The other option for using MicroSort® is in vitro fertilization (IVF). This is where your egg is removed in a process called an egg retrieval. This is a detailed process that begins with a close monitoring of your cycle.

Many medications, including injectable ones, are used to chemically control your cycle. The process of ovulation is managed and your body is chemically encouraged to produce many more eggs than you would normally produce. Once the eggs reach a certain stage, as predicted by an ultrasound exam, you will take another medication to prepare them for ovulation and retrieval.

The process of retrieval usually involves general anesthesia or intravenous sedation to place you in a very sleepy and comfortable state. A long ultrasound-guided needle is placed through the vagina and used to remove eggs from the follicles in your ovaries.

After the eggs are removed, you are awakened from anesthesia and allowed to recover from the effects of the medication. You may be a bit groggy for up to 24 hours, depending on the medications given. Over-the-counter anti-inflammatories, like ibuprofen, are usually more than sufficient to deal with any mild cramping you may experience.

In the meantime the sperm sample has already been sorted as either an XSORT® or a YSORT®. It will be mixed with your eggs in a petri dish at the lab. The dish is placed in an incubator and watched for development.

You will then wait for three to five days to do the embryo transfer, depending on the protocol of your clinic and other factors. The procedure is usually quick and not very painful. A small catheter is threaded through your cervix. The embryos are deposited inside your uterus. Typically you will be asked to lie still for a few minutes to a half hour. Then you are released to await a pregnancy test in 10 to 14 days.

One of the most convenient options, when using in vitro fertilization, is that you can choose to do your MicroSort® cycle at the Genetics & IVF Institute in Virginia, at the Huntington Reproductive Center in California, or with a local collaborating physician. MicroSort® has a list of current collaborators on its website. You are free to chose any of them, though each one's fees and cycle rules may differ from the others. It is important to point out that if your personal reproductive endocrinologist is not listed as a current MicroSort® collaborator, he or she can apply to become one. Says Dr. Karabinus:

> "Collaborating physicians must enter into a collaboration agreement with GIVF in order that they may receive MicroSort® sperm for their patients. The agreement spells out the responsibilities of being a collaborator who may participate in the clinical management of the trial participant. The collaborator agrees that the MicroSort® semen may not be shipped to other than GIVF or other collaborators. The collaborator agrees to provide information regarding the cycle in which the MicroSort® semen was used. The collaborator must have an IVF laboratory on the premises or be affiliated with an IVF laboratory."

Using your own physician could be a real possibility for you. This may even save some travel time and expense. To enroll as a collaborator, your doctor will need to call Genetics & IVF Institute to begin the process; you can't be the one to do it.

As Dr. Karabinus points out:

> "50 to 60 percent of MicroSort® clinical trial participants come to GIVF for treatment. The remainder participate through their collaborators."

This gives your local physician some idea of how many other physicians are performing and profiting from MicroSort® services outside the Genetics & IVF Institute location. This might be a selling point if you are having trouble getting your physician to agree to become a collaborator.

Monitoring for Ovulation

Your ovulation is monitored in multiple ways and can be done in numerous locations. You can choose to have your ovulation monitored in your hometown by your local physician. This assures you the close medical support you need at a local level. Local monitoring with your collaborating physician is only an option, though, if you choose to pay the extra fees.

You may also be monitored for ovulation at the facilities where you will have your sort. This can be at the Genetics & IVF Institute in Virginia or the Huntington Reproductive Center in California. Once again, the fees for monitoring are in addition to MicroSort® services. This can add travel and hotel/accommodation fees. Most patients do not have the time, energy, or money to spend away from home or their jobs for a process this lengthy.

Many women choose the least-expensive route: home monitoring with the use of ovulation prediction kits. Since these kits are fairly self-explanatory and easy to use, most families have no issues with it. Should you desire some extra reassurance, there is also a limited package of monitoring available from Genetics & IVF Institute for a small fee.

Other Options for In Vitro Fertilization

If your husband has male factor infertility problems, an additional option when doing IVF is the use of intracytoplasmic sperm injection (ICSI, pronounced "ick-see"). This is where a single sperm is injected into a single egg. This provides another way to increase the odds in your favor of selecting the desired sex of your baby. Intracytoplasmic sperm injection with sorted sperm definitely increases the likelihood of your conceiving a child of the sex you desire.

Another option, which we will discuss in more detail in the next chapter, is preimplantation genetic diagnosis (PGD). This can also be done in combination with MicroSort®. Preimplantation genetic diagnosis (PGD) is available at both the Genetics & IVF Institute and the Huntington Reproductive Center.

Additional options include gestational surrogates, as well as egg and sperm donors. Both centers are readily available to discuss the variety of options available to you and your spouse. They have been in business for many years and have dealt with a number of situations, so they likely will have solutions you have not thought of, making them an invaluable resource in your quest for parenthood.

Getting Started with MicroSort®

The staff and physicians at MicroSort® are extremely friendly. I say this because it is very important that you are able to get your questions answered in a prompt, clear, and polite manner. Given the personal nature of some of the questions you or they may be asking, you will want to know that you and your private information are being handled with the utmost care.

One of the things that truly impresses many patients is that a real person answers the phone when you call. Says one MicroSort®-hopeful mom:

> "Even when you're not sure of the exact questions you want to ask, everyone I have ever spoken to was polite and friendly. I never had to drill down through one of those annoying answering systems that give you 12 extensions. I always had a knowledgeable, pleasant, live human being. This always made asking questions very comfortable."

"Patient care at GIVF has always been foremost at GIVF," Dr. Karabinus proudly relates. "We have worked hard over the years to provide the best possible care to all patients who come to GIVF."

The decision to undergo sex selection is not an easy one to make for any family. While you are trying to make that decision, you should check out the MicroSort® website for the most current information. It offers a wealth of information on the practical aspects of sex selection therapy at Genetics & IVF Institute as well as a section for frequently asked questions about the entire process.

After many discussions, some soul searching, and a look at other issues that will factor in, you may decide that MicroSort® is the best option for your family. Unfortunately, because of the clinical trials, not every family is eligible to be treated.

You must meet certain specific conditions to qualify for the MicroSort® trial currently under way. These MicroSort® requirements include:

- Having at least one child already
- Being a married couple
- Mother being between 18 and 39 years old
- Both father and mother testing negative for HIV, hepatitis B, and hepatitis C
- Sorting only for the underrepresented sex in the family
- Meeting other criteria for pregnancy and children's health history

Since the MicroSort® procedure is currently undergoing clinical trials, the researchers are quite selective about whom they accept. This has led some women to think that certain other women out there will lie or be devious to get accepted into the program. MicroSort®, however, has put into place every safeguard to prevent this from happening, so as to protect their data from being polluted. In the end, such caution should be genuinely helpful to everyone who wants to do MicroSort®, even if it occurs down the road after the clinical trials have ended.

If it turns out that you are eligible, your next step is to return to the MicroSort® website. There you'll also find the most up-to-date paperwork. Start by printing off the packet of information and forms from the website. Read every thing very carefully. Develop any questions you have about the method; be sure to write them down so that you do not forget them.

The paperwork is lengthy and must be filled out meticulously. Because the process is now in clinical trials, participation in it is rigorously governed, so as to protect the integrity of its results; in turn, this protects you, the consumer. So follow all the instructions closely. For example, there is one form that you are to read but not fill out until after your consultation.

You will fill out medical histories for both partners. You will complete consent forms for the release of medical records on you, your spouse, your past pregnancies (if any), and your children's developmental histories. You will also find information about testing for certain medical diseases like HIV and hepatitis.

Once you have filled out the packet and obtained the necessary paperwork, you will set up an appointment with a consultant from MicroSort®. This can be done via the phone or in person at either of the locations in Virginia and California. There will be an interview where you will discuss your medical history, be given the details of your options, make certain choices about your cycle, and be told what your chances are—not only of conceiving, but of conceiving a boy or a girl (depending on which sort you intend to do). Your consultant can help you decide if this is the appropriate step for you.

MicroSort® Fees

How much will it cost? This is a common question. The sex selection business is pretty much a cash and credit business. Only in rare instances will your health insurance cover the fees associated with sex selection.

Some companies will cover a few of the prequalifying screenings, like the HIV and hepatitis testing. You may also want to check with the administrator of your Flexible Spending Account (FSA), if you have one on your job, to see if you can use money from it to help pay for any or all of your cycles.

The fees that you will be quoted by MicroSort® include only the provider's fees for services delivered as part of the MicroSort® procedure. The quotes do not include fees from collaborators, medications during your cycle, and fees for other potential outside laboratory services or medications. You can usually get an estimate written up ahead of time for fees like medications. (All fees cited below were current as of September 2006; please check the MicroSort® website, at *www.microsort.net* for fee updates.)

All fees will need to be paid in full by day one of your MicroSort® cycle. This payment can be from various sources, including your bank account, credit card, line of credit, flexible spending account, and the like.

The first cost you will incur with MicroSort® is the $300 consultation fee. This is required of everyone. It does not matter where you intend to do the sort or how you conduct the interview—by phone or in person. This fee is nonrefundable, even if you decide that MicroSort® is not for you or the MicroSort® representative decides you are not right for them.

The Genetics & IVF Institute charges $3,765 for a cycle and sort in which you use home monitoring. In addition to this charge you will need to factor in the costs of monitoring either at home, at your local physician's office, or at a MicroSort® facility. Your travel costs will depend on how far you have to go and how long you must stay.

Sometimes you may not feel overly confident in your ability to predict ovulation. The very thought of throwing good money down the drain based on your own predictions of ovulation can be scary. After all, don't you need the initials "M.D." after your name to make it all valid? GIVF actually offers a small package of monitoring. For a maximum of two days, they will monitor you to confirm what your ovulation prediction kits are showing. This only costs $475. Some moms report that choosing this option made them feel more hopeful.

Some families have decided to have all their monitoring for the cycle done at Genetics & IVF Institute. The cost of $5,415 includes the sorting of sperm, the intrauterine insemination, and the daily monitoring. For families who do not live close to the centers, the additional cost of travel and accommodations may make monitoring at a MicroSort® facility unreasonable, for either time constraints or financial concerns.

Monitoring at your local reproductive endocrinologist's office is available if you live a good distance from Genetics & IVF Institute. This is known as outside monitoring. Because of the collaboration needed between physicians, you will be charged by both GIVF and your local physician. This brings the GIVF fee of an outside monitoring cycle to $4,390. That includes the sort, the intrauterine insemination (IUI), and the collaboration. You will also need to cover the costs of the laboratory testing and the fees of the collaborating physician, which are separate.

Currently the cost to sort one vial of sperm is $3,400. If you have a semen sample that is large enough and of high enough quality, you may be able to get a second vial out of that sample. There is an additional fee of $700 for the additional vial, if the sample is done at one of the GIVF centers; otherwise, the fee does not include shipping, freezing, or storing the vial.

There is also a holiday or weekend penalty. If you require the sorting procedure to be done on the weekends, you will pay a premium of an additional 10 percent of the sort fee. That number goes up to an additional 20 percent, if you require their services on a holiday.

Have you ever heard that improper planning will cost you? This is also true at MicroSort®. You will have an additional 20 percent fee tacked onto your bill if you require a sort with less than seven days' notice.

Fortunately, certain financial assistance plans are in place for qualified X-linked-disorder participants in the MicroSort® trials. This benefit may reduce or even waive the fees surrounding the actual sort. But other fees that may be incurred, such as medications, travel, consultation, and so forth, still apply.

Why MicroSort® Might Be Right for You

MicroSort® delivers many benefits in the sex selection arena. This is one of the reasons that this method of sex selection is becoming popular with families who are choosing to do sex selection for a variety of reasons. One of its biggest benefits is the stunning success rates it has achieved. This is particularly true when you look at the low-tech methods of sex selection. MicroSort® has the science and numbers to back up its success record. Scientists agree that these numbers are impressive and accurate. Even the most popular sex selection methods that are low-tech don't have the backing of scientific evidence that MicroSort® has amassed behind it.

The success rates are only one of the multitude of reasons that many families choose to use it. One mother who was shopping for a method of sex selection told me that she was leaning toward MicroSort® largely because of its great number of successes.

> "I could try a couple of practice cycles with other methods. I could spend a lot of time trying to figure out exactly when I ovulate. Then I can take my chances on my own and get only slightly greater than 50/50. Even if I had the money to have an unlimited number of children, which I don't, I simply do not have the time. I really feel my biological clock ticking at 37. And I want a daughter, like, yesterday."

Another popular reason to choose MicroSort® is that you get high success rates without having to do in vitro fertilization. While in vitro fertilization is an option for some, it is not necessarily a boon for all. It can also add a certain level of physical and emotional stress that not every family is willing to undergo for sex selection, in addition to the monetary outlay. This is a personal decision that only you and your family can make.

Since the basic MicroSort® process has been used for many years in the agriculture industry, the system's kinks have been worked out. In fact, the process is still used today in agriculture. In the process of applying for the clinical trials, MicroSort® even had to prove that there was no harm to future generations. These tests were carried out on generations of quickly reproducing species—like rabbits. In such tests the scientists breed multiple successive generations after the MicroSort® procedure and then look for problems in future generations. None has been found in up to five generations of animals born from the MicroSort® process.

For those who choose to do in vitro fertilization or preimplantation genetic diagnosis, MicroSort® can help them sway the odds in their favor. It is possible that a couple would go to the trouble of doing preimplantation genetic diagnosis and then find out that none of their embryos was of the sex they wanted. But by using MicroSort®, they greatly improve their odds of having embryos of the desired sex.

Potential Drawbacks to MicroSort®

It would be unfair to give you all this great information about the amazing technology that is known as MicroSort® without also discussing some of its potential drawbacks. As with anything, what may be seen as a drawback to some is seen as a benefit to others. There are also some benefits, as well as limitations, that won't apply to you or your case. All this information should be discussed with your practitioner.

MicroSort® is not 100 percent effective. Some would argue that its relatively low cost and the lack of need for everyone to do in vitro fertilization makes it the best bargain in sex selection. Nevertheless, it is probably *not* the best option for you if ending up with a baby of the opposite sex is something you absolutely cannot handle. Yet MicroSort® might be a useful addition to preimplantation genetic diagnosis for you, if you feel that strongly.

Another potential drawback might be the pregnancy rates. Even with a perfectly healthy couple with normal fertility, the clinical pregnancy rate is about 15.4 percent per treatment cycle with MicroSort®. If you choose the IVF/ICSIS route, that pregnancy rate goes up to 35.7 percent. *(Data accurate as of November 2006 per microsort.net.)*

The MicroSort® procedure done by using intrauterine inseminations achieves lower pregnancy rates than by using in vitro fertilization. This is true even for women of normal fertility. You should therefore plan to do multiple cycles to achieve a pregnancy of any type. Of course, some couples conceive on the first try, but it is best that you not count on this possibility.

There is also the fact that MicroSort® is still in the clinical trial phase. You must consider the possibility that something will be found that discontinues this process, though this is not considered likely in the case of MicroSort®. More than 850 babies have been born already, with more planned.

Some people also wonder about unforetold complications. Will the use of dyes and lasers produce problems in the offspring years later? Perhaps even as they are trying to have children of their own? The answer is that we currently do not know.

Combining MicroSort® with Preimplantation Genetic Diagnosis (PGD)

When looking at the high-tech options of MicroSort® and preimplantation genetic diagnosis (PGD), you might be wondering why you would bother to even think about combining the two processes. While you might cite expense as a reason to forgo adding both preimplantation genetic diagnosis and a round (or more) of in vitro fertilization to the cycle, there is a big benefit to be gained.

The benefit is that if you use MicroSort® before PGD, you actually increase the likelihood that you will have more embryos of the sex you desire. It is technically possible to be doing PGD and everything else that it entails, and then wind up with embryos that are all of the "wrong" sex. But by using MicroSort® you increase the chances that this won't happen.

9

Using Preimplantation Genetic Diagnosis

History

In 1967, when Robert Edwards and David Gardner successfully determined the sex of rabbit embryos, the science of assisted reproduction had reached a point from which there was no turning back. The chain of events that followed led to the development and refinement of both in vitro fertilization (IVF) and preimplantation genetic diagnosis (PGD). It wasn't until the late 1970s, though, that a baby was born from the use of in vitro fertilization—the legendary Louise Brown in England.

Dr. Mark Hughes is credited with having discovered preimplantation genetic diagnosis, though in those years the technique was still very much in its infancy. In 1990 the first children were born from embryos after undergoing PGD. These first cases were used to prevent X-linked disorders.

Today the process of preimplantation genetic diagnosis has been greatly refined. The procedure is available and ready to use for those few patients who need PGD to help prevent their children from developing specific genetic diseases. The use of preimplantation genetic diagnosis for sex selection is still fairly controversial in some parts of the country, despite its ultrahigh, nearly perfect success rates in selecting a baby of *either* sex.

One reason for the growing popularity of preimplantation genetic diagnosis is that it is nearly 100 percent effective at finding the genetic disorders for which it tests. This testing is also nearly 100 percent effective in sex selection, even for boys. Overall, then, this is the most accurate form of sex selection available.

Basics of Preimplantation Genetic Diagnosis

Preimplantation genetic diagnosis is always done jointly with in vitro fertilization, though it is not always done for fertility issues alone. Many couples with no history of fertility issues are agreeing to do in vitro for a variety of reasons, including sex selection.

Since using preimplantation genetic diagnosis means also using in vitro fertilization, couples must be willing to spend the extra money that it will cost to do a cycle. Still, the benefits are numerous, even in the sex selection route. This is because using preimplantation genetic diagnosis is extremely accurate for either sex. It also boasts a higher pregnancy rate per cycle than many of the other high-tech methods of sex selection, despite its lofty price tag.

Embryo Biopsy

The big deal with preimplantation genetic diagnosis is the embryo biopsy. This is the portion of the process that actually pinpoints the sex selection. A couple of different types of embryo biopsy are done, though the basics of each are about the same as far as you, the user, are concerned.

For most of these procedures, embryos are allowed to grow in an incubator for a couple of days, dividing away, blissfully ignorant of what's to come. The first stage of this process is to select the best-looking, healthiest embryos that have been growing. At this point there may not be as many as you started with.

The embryologist then uses a tiny, hollow, glass needle to extract a single cell from the embryo. This is carefully done with great skill using a microscope and gentle suction to hold the embryo still during the biopsy.

At this stage of development the cells have not yet differentiated, meaning that removal of the cell will not harm the embryo in any way. Some studies have shown that you can remove up to 25 percent of the embryo without harming it, though typically only 1 of the 6 to 10 cells is removed.

The embryo is then returned to the incubator, while the extracted cell is kept separately prior to testing.

Polar Body Biopsy

In contrast to an embryo biopsy, a polar body biopsy can be done even before conception. This method is more acceptable to some because there is no risk to the embryo. It's able to tell if there is something wrong with the oocyte, or mother's contribution, but it can't evaluate the male contribution. It is, therefore, not highly relevant in terms of sex selection, since the chromosome that determines the sex comes from the male.

Blastomere Biopsy (Cleavage-Stage Biopsy)

The benefits of a blastomere biopsy is that it can provide information on the genetic makeup of both contributions of the father and the mother, as well as tell the sex of the embryo. The ideal timing for this type of biopsy is believed to be at the eight-cell stage. That stage also provides enough time for the embryo to be returned to the uterus at the appropriate moment. The main question is whether the information gathered at this stage is representative of the entire embryo. It has been reported that 94 percent of all preimplantation genetic diagnoses use a blastomere biopsy.

Blastocyst Biopsy

This method of embryo biopsy is considered to be the best in terms of being able to find genetic issues in either the maternal or the paternal contribution. It also works perfectly for sex selection. One of the big issues is the amount of time it takes to run this type of biopsy. There is less time to double-check before needing to return the embryos to the uterus. The other issue is that many of the embryos don't survive long enough to be available for this process. Therefore, it can be a double-edged sword.

Three Genetic Tests Done with Preimplantation Genetic Diagnosis

Preimplantation genetic diagnosis has the benefit of offering more than a simple screening of the sex of your baby. In fact, three main types of tests are done during the preimplantation genetic diagnosis process:

Sex determination—This is merely checking for the X or Y chromosome to determine whether the embryo in question is a boy or a girl.

Chromosome abnormality (also known as aneuploidy)—This screening is done to ensure that the embryo being tested has the correct number of chromosomes. Aneuploidy is the state of having the wrong number of chromosomes. This could happen when there are either too many or too few chromosomes.

Single gene disorder—This is the search for a specific known disorder that affects a single gene. The embryo can be tested for actually having the disorder in question, as well as for whether the embryo would carry that disorder. Currently there are many diseases that can be screened for in the single gene category.

Screening for these additional embryonic issues can actually increase the number of successful pregnancies. This is because you are starting with only the best embryos from a given cycle.

This can lower the rates of miscarriage and other pregnancy losses from chromosomal abnormalities.

Intracytoplasmic Sperm Injection

Discovered in 1992, intracytoplasmic sperm injection (ICSI, pronounced "ick-see") is the placement of a single sperm inside an egg for the purpose of fertilizing that egg. Originally intracytoplasmic sperm injection was used for severe male factor infertility, though now the reasons for using this new technology are varied. Because only a single sperm was needed, even men who had incredibly low sperm counts (or other male factors) were able to father children.

The process starts by collecting a semen sample from the male and retrieving eggs from the female. The semen sample is prepared by washing it. The embryologist will prepare the egg by removing the cumulus and coronal layers from the egg. With intracytoplasmic sperm injection, single sperm are pulled out from the sperm sample. Then using a very small, hollow, glass needle, that single sperm is directly placed into an egg. This fertilized egg is now ready for the incubator, where it will await preimplantation genetic diagnosis at the appropriate time.

The fertilization rate for most fertility centers using intracytoplasmic sperm injection is 60–85 percent.

Sperm can be retrieved in small numbers directly from the testicles if necessary. This can be done even in men suffering from obstructive azoospermia, where no sperm are found in the epididymis because of a blockage. The retrieval is done with a fine needle biopsy, called a testicular sperm aspiration (TESA). It can be done under local anesthesia.

In the testicular sperm aspiration, testicular tissue is removed and placed into a special medium to protect it. In the lab this tissue is processed to remove the sperm from the surrounding tissues.

If your partner suffers from nonobstructive azoospermia, the procedure is different and less successful. You will need to have him undergo something called a testicular sperm extraction (TESE). This form of infertility means that not much sperm is present to harvest. So the TESE is used to biopsy multiple sites in hopes of finding more sperm. While this can be done under local anesthesia, some form of sedation is frequently used. Any sperm retrieved can be frozen for future use.

You should consider intracytoplasmic sperm injection if:

- You have severe male factor infertility
- You don't want to use donor sperm
- Sperm motility is less than 35 percent
- Poor sperm morphology is found
- Low sperm concentrations are noted (less than 20 million per mL)
- Your previous cycle had little or no fertilization
- A low number of eggs is retrieved

Some families choose to use ICSI even when there is no male factor infertility to deal with. They feel that this gives them the advantage of choosing only well-formed sperm. This is usually not an issue for the fertility clinic that you are using, though it will charge extra for the procedure.

Some worry that the micromanipulation of the egg and sperm can lead to potential problems with the baby. To date, numerous studies have been conducted that have found no additional risk of either genetic defects or birth defects to intracytoplasmic sperm injection offspring. The most compelling evidence is the more than 100,000 babies born after use of this technique.

Using this technology in addition to the other costs is not inexpensive. In fact, it can cost upward of $2,000, in addition to the other fees associated with in vitro fertilization and sex selection.

Using Preimplantation Genetic Diagnosis with MicroSort®

Many couples who are choosing to do preimplantation genetic diagnosis for sex selection are also combining it with MicroSort®. The biggest benefit to doing both is that this will increase the number of embryos of the desired sex.

Deciding on doing both can be a difficult decision for many families. It adds extra steps and an increased fee. It can involve more red tape. It might even be that you do not qualify for the MicroSort® trials. This can leave you wondering what to do, should you find that all the embryos you have are of the opposite sex from the one you wish.

After all, it is possible to go through the entire in vitro fertilization and preimplantation genetic diagnosis process only to find that you have created zero embryos of the sex you desire. This leaves you with the unenviable choice of having to scrap the entire cycle or to put back healthy embryos, but of the "wrong" sex.

Considering the amount of money to be spent on in vitro along with the addition of preimplantation genetic diagnosis, the fees for MicroSort® are a drop in the bucket, so to speak. Many couples therefore decide that using MicroSort® adds insurance to a lengthy, emotional, and costly process.

The Problem with Preimplantation Genetic Diagnosis

While many do not debate the benefits of using preimplantation genetic diagnosis to prevent single gene disorders and chromosomal disorders, its use for sex selection is highly debated. This leads to a discussion of the problems associated with the technique, both in general and as it relates to sex selection.

Preimplantation genetic diagnosis has roughly a 5 percent false abnormal rate. This means that an embryo is described as abnormal, when in fact it was perfectly normal. The false normal rate is about 2 percent, meaning that an embryo is identified as being free of disease and chromosomal disorder when in truth it's abnormal. There is also the problem that human error can always be involved. The chance of error for sex selection alone is all based on human error, so PGD testing is nearly 100 percent effective for sex selection.

Embryos die in preimplantation genetic diagnosis. This means that the cell biopsy can cause an embryo to cease to exist. For some, this presents an ethical dilemma. The dilemmas with PGD, however, go far beyond the death of an embryo after a cell biopsy.

Dr. Mark Perloe, a reproductive endocrinologist, suggests that:

> "With experienced technologists, the risk to the embryo is minimized and the potential to pinpoint the best embryos is at its best. There are 11 other pairs of chromosomes that are checked at the same time."

Then you have to ponder issues of what to do with the embryos of the "wrong" sex. Do you donate these embryos to science? Do you donate them to other infertile couples? Or do you choose to save them for later? Other couples just wind up with the embryos in a state of frozen anticipation while they try to decide what to do.

Dr. Mark Hughes, the father of PGD, is not in favor of using it for sex selection. He designed the technique to help prevent disease and to bring children to couples who might not be able to pay the emotional price of conceiving a child destined to be ill or die.

Questions to Ask about Preimplantation Genetic Diagnosis

- Which type of preimplantation genetic diagnosis do you do?
- When do you make the embryo transfers?
- Do you insist on or recommend that we do ICSI?
- What do I do with the left-over embryos?
- What do I do with the embryos that are not of the sex that I originally chose?

- Do you prefer using PCR (polymerase chain reaction) over FISH (fluorescence in situ hybridization)?
- Do you offer an in vitro guarantee? Will I still get a guarantee if you do a PGD?

How the Process Works

Once you and your partner have decided that preimplantation genetic diagnosis is the route for you, then you will begin the process. The first step is to find an infertility clinic where you can do you in vitro fertilization cycle. Choosing the appropriate center can be more difficult than you might think. You need to make sure that the fertility center you are using has the capability to do preimplantation genetic diagnosis and is also willing to do it for sex selection only.

Some fertility centers offer preimplantation genetic diagnosis but not on site. This means that you will have the added cost of having an alternative source do the cell biopsy. This can happen in a couple of ways. Some centers have you pay to fly the embryologist and his or her equipment in for the cycle. Other centers will do the embryo biopsy and ship the cell to the embryology lab.

If this is not acceptable to you then you will need to find a center that actually does preimplantation genetic diagnosis in house. This might mean that you will need to travel to a center that is not conveniently located. Few cities have the ability to sustain multiple fertility centers, thus limiting your choices.

You will want to make an appointment as soon as you can. Unfortunately, the wait times for even a consultation appointment can be very long, up to several months. This is a horribly long wait when you are eager to get started. When you make the appointment, ask what you will need to bring and if there is anything you can do before the appointment to make the process go any faster. You might want to ask whether you can do any semen analysis or infectious disease testing before the consultation to save time.

They may explain the details of fertility charting to help you learn your cycles and pinpoint your ovulation. They may also recommend any available books or classes. I highly recommend that you do all these things, even if you consider yourself an expert on the topic. Not only will classes give you something to do while waiting, but in the absence of new information they'll also enable you to learn how your practitioner thinks and works.

When you report for your consultation, be sure to bring all your questions written out. This will help you remember to get them all answered. You might also show up with as many months of charts as you have available. This can give you a head start on the process.

Once you have the consultation with the clinic doctor, you will find out what you need to do before the cycle can begin. Many fertility centers will require that you do infectious disease testing. This can include tests for HIV, hepatitis B, hepatitis C, and others. You may also be asked to do a psychological evaluation, though not every clinic will ask you to do this step.

You may need to do a preparatory cycle, which consists of taking various medications to prepare your body for the controlled hyperstimulation of ovulation induction. Your doctor can tell you if this is required, depending on his or her protocol as well as your fertility status.

Then you will receive information on the medications you can expect to use and how the monitoring will work. During the medication phase, you'll have repeated ultrasounds and blood work to test your body's response to the medications. Your meds can be adjusted either upward or downward as needed. The goal is to produce a number of really good eggs during the cycle.

During this part of the cycle, your husband or partner has very little to do. He should try to help support you mentally and emotionally. This can be an emotionally raw time for both of you.

You will be told when to take your shot of human chorionic gonadotropin (hCG) to trigger ovulation. About 36 hours after you trigger, you will have your eggs retrieved. This is typically done at the fertility center.

When you report to the fertility center for your egg retrieval, you'll also bring your partner. He will be asked to produce his semen sample during this time. Some fertility centers allow samples to come from home. You simply need to ask if this is possible.

While the preferred method is through masturbation, there's occasionally an exception made for intercourse with a special collection condom. Your fertility center may or may not have this available as an option. You shouldn't use oral sex as a collection method because saliva can harm the sperm, making your sperm sample useless.

You will then be prepped for the retrieval. This will include changing into a gown. You will have your vital signs taken and an IV will be inserted. That's done because this is technically a minor surgery that usually requires general anesthesia or conscious sedation, which will render you unaware of the procedure.

You will be taken into the retrieval room. The anesthesiologist will render you sleepy so that the reproductive endocrinologist can retrieve your eggs. The process involves an ultrasound-guided needle. This needle is long enough to go through your vagina to your ovaries to remove the eggs. The ultrasound is used to ensure that all the follicles are tapped. The process is done on both sides, so that each ovary is emptied of mature eggs.

The eggs are then handed over to the embryologist in the lab. Usually the lab is one window away from the retrieval room, for both proximity and efficiency. You will be awakened and taken to a recovery area to wake up a bit more. Here you can join your partner. You should know before you leave how many eggs were collected from your retrieval.

After an hour or so, you are sent home with instructions to rest. Typically you can expect light cramping and some spotting.

You're encouraged to rest and refrain from driving due to the anesthesia effects more than anything else.

Once handed over, your eggs will begin a process with the embryologist in charge. The sperm is washed and prepared for the creation of embryos. The eggs and sperm are put together into a petri dish. The embryologist checks periodically to see whether conception has occurred.

If you chose to go the intracytoplasmic sperm injection (ICSI) route, it will be carried out as soon as the sperm and eggs are prepared. The fertilized eggs will then be placed in an incubator until it is time for the embryo biopsy.

How MicroSort® Changes the Process

The process is generally the same when the MicroSort® is added. You will want to contact the MicroSort® clinic to see about applying through them as well. Since they're doing clinical trials, you'll need to be able to meet certain criteria to qualify for this process. For more information on these criteria, please see Chapter 8, dedicated to MicroSort®.

Since you have decided to use the preimplantation genetic diagnosis route for sex selection, you will be able to use the frozen sperm option with MicroSort®. This means that you won't have to travel to either of the facilities (Genetics & IVF Institute or Huntington Reproductive Center), but instead can just ship your semen sample. It will be sorted at the facility you have chosen.

Once they have done the XSORT® or YSORT®, your sample will be sent back to your fertility center and stored, frozen, until your cycle. This should be done well in advance, otherwise there are added fees associated with a rush cycle.

The combination with MicroSort® enhances the process in many ways. The number one reason that parents use MicroSort® along with preimplantation genetic diagnosis is typically to select the sex of their baby. Physicians feel differently and offer up more scientific rationales. "For medical reasons, backing up MicroSort® with preimplantation genetic diagnosis results in very few errors," says Dr. Perloe.

Preimplantation Genetic Haplotyping

The technique known as preimplantation genetic haplotyping (PGH) varies from preimplantation genetic diagnosis in that it can screen for multiple disorders, both known and unknown. The DNA extracted from the embryo is amplified, which means that the results are less likely to be faulty. The family also undergoes screening. Other than this, the process for the embryo biopsy and everything else are similar.

The benefit of PGH is not sex selection alone. The true benefit is for parents who have concerns about single-gene issues and aneuploidy. These parents can also use this new technology for the purpose of sex selection.

Mark Robinson, a molecular geneticist at London Bridge Fertility, Gynaecology, and Genetics Centre, offers this comment:

> "PGD involves testing for single gene disorders and this can be performed with either a combination of direct detection of the mutation in conjunction with linked markers or in some instances with linked markers alone. Where PGH differs is that it relies solely on the use of linked markers. Due to the inefficiency of the method used to produce the tested material, in general, more markers are required for PGH than PGD and it is the use of several markers that produces a haplotype."

PART FOUR

Making and Implementing a Decision about Sex Selection

10

Talking to Others about Sex Selection

Many of the discussions in our lives are difficult—particularly when they revolve around matters of the heart, like the matter of sex selection. The discussion of future children may be one of these charged topics, particularly if you are bringing up the subject of selecting the sex you would like them to be.

You may feel confused and unsure of what to say to friends and family. You may not have a good handle on how you are feeling. The words may not flow as eloquently from your mouth as you had hoped. Altogether, it can be a very awkward conversation to have, even when conducted with the best person in the world to have this conversation with—your spouse or partner.

Figuring out What to Say

Before you can talk to others, even those especially close to you, you need to know what it is that you intend to convey. What are the major points you would like to get across to the other person? Is it your long-held desire for a baby boy or a girl? Is it the need you feel to experience the raising of that son or daughter you have always dreamed of? When strangers talk about sex selection as if it is something out of Dr. Frankenstein's laboratory, do you simply want to stand up and scream that they don't know what they are talking about? That they are discussing real live, loved babies?

You might find, however, that there are people out there who think and feel as you do. Perhaps they too are keeping quiet for fear that they are the only ones around having the same thoughts about the possibility of sex selection in their lives. Maybe they have feelings quite similar to the ones you are sharing.

Opening yourself up in this discussion is difficult, whether it involves talking to your spouse in the privacy of your home or having a heart-to-heart with a friend who hasn't known about your desire in this area. Here are some ideas to keep in mind when preparing to talk to other people:

- List your ideas on paper, even if you don't write anything but a couple of words or short sentences.
- Write down your goals for the outcome of treatment. Do you hope to have a baby of a specific sex? To have that baby by using a specific technology?
- Find news or science items that would appeal to the person you are trying to talk to about sex selection.

Sometimes you do not need someone to talk to you, you just need a good friend or your spouse to listen to you. There is something about the validation of your feelings that can help you feel less "heavy" about the issue of sex selection. Being able to convey your needs to those around you can result in positive effects on your anxiety level.

Talking to Your Spouse

When you have something of this magnitude to discuss, it is recommended that you think about the conversation before you actually have it with your spouse or partner. Write down a couple of talking points that you want to make when discussing this topic. Think about the questions that you anticipate from him; this helps you plan your responses carefully.

JoJo says that her discussion actually started at her ultrasound in her second pregnancy. And in the end it was her husband who brought it up.

> "I had the whole day planned and knew we would end that day with going to this store where they had the most adorable Cinderella dress. When the tech said it was a boy...the look on my face said it all, and all he could do was feel sorry for me. Later that night he said, 'Hey, next time we can look into some sperm spinning or something.'"

That was the start of a discussion between the two of them. JoJo's husband was equally supportive when she did a sex test that revealed the opposite of what she was hoping for. Says JoJo:

> "With this pregnancy I did research ahead of time and knew I would get a girl. I am still waiting on the confirmation that it is indeed a boy, but when my gender test came back and all I did was cry for days and days, I didn't have to approach my husband. He said, 'Next time, IVF or MicroSort®.' I think most of our husbands just know the longing in our hearts for a particular gender. This is the easiest person to convince, in my opinion."

One mother estimated that for the couples she knows, 80 percent of the time it's the woman initiating the discussion of sex selection. The Internet forums and chat rooms are certainly buzzing with discussions of women telling their husbands and getting them to agree. You might be surprised, though, at how many fathers out there are also big believers in sex selection.

Kevin says that his desire for a daughter was all-consuming. His family was blessed with two healthy and lovable sons, but he wanted a girl. After reading up on a website about the various technologies that he and his wife could use to conceive a baby girl, he knew he wanted to try for a third baby.

He wondered if it made him a "creep" to ask his wife to go through all these invasive medical procedures just so that he could have a daughter. He says:

"It wasn't like she needed to do in vitro or anything. So here I was, asking her to get daily shots to have a baby girl that I wanted. She was happy with the boys; I was the one who wanted more. In the end I knew it would be worth the trouble she went through, but how could I convince her? Or even bring it up to her in the first place?"

He laughs as he recalls his words that fateful day when he told his wife:

"'I'd love to have another child. And if there were a way to improve our chances of having a girl, would you be willing to try? It would mean a lot to me. It's something that I would really want to try.'"

Luckily for him, she accepted the offer with little prodding. Now that Kevin has his little girl, he's thrilled. His wife is pretty happy too.

"She never really thought about being the mother of a girl, but I think she's really gotten into it. Our house is so happy. All four of us fight over who gets to hold the baby—even the boys."

What to Do If You Don't Agree with Your Spouse

Unfortunately not every conversation with your spouse will go as you have planned. At times, the two of you simply will not agree on the subject. Your disagreement might be something minor that discussions and negotiations will resolve. But it may turn into a conflict of a greater nature.

Taking your time to resolve all the issues and conflicts is the best way to approach this type of reaction. Many women say that they are the ones who initiated the conversation.

If you find that your spouse does not want to do sex selection, try to have conversations to figure out why he feels that way. Is he opposed to sex selection in general? Does he prefer a baby of the opposite sex? Or are the financial or physical issues involved with high-tech sex selection just too much for him to deal with?

Mary is the mother of four—three boys, plus a daughter born after three cycles of MicroSort®. She says that it was all her decision, and that her husband felt very satisfied with every boy. Mary says:

> "I tried to get over it and hoped it would go away but it didn't. I always wanted a girl."

Eventually she told her husband:

> "'I'm 35. Now is our chance—let's do it.' And he finally agreed. This was 100 percent me. He did this for me. Now that she's here, he's so close to her. Oh my goodness, she's so different—she's a girl!"

Figuring out Whom to Talk to in Real Life

Talk to people who are supportive. It might sound like common sense, but sometimes those who support you are either hard to find or hard to understand. Remember that a bit of kindness goes a long way.

You will find others who feel similarly, and you can just talk to them. They may be people you have known forever, or they may be new friends. They may be people who are sitting next to you at the fertility clinic or ones that you meet in play groups.

There will be people who say things that you do not like about sex selection or about parents who choose to do sex selection. These people may not even know that you are considering sex selection or are actively in the process of doing it. This is probably more ignorance of the topic than anything else.

The technologies are new and many people are understandably worried about them. Does the opportunity to pick the sex of your baby make this too much like *Brave New World* for some? You bet! The fact is that you do not need to let their ignorance hurt you or your feelings. Chalk it up to what it is—usually fear and misunderstanding—and move on with your life. Ignore those who have nothing nice to say; it's not their life nor their decision.

Ignoring your friends and family is not possible. Doing so would make the situation more tense and difficult, according to JoJo. "Let's just say in my case I have been labeled a horrible mother. I was told that I was unloving and ungrateful and willing to play God. People who imply such things obviously have both sexes so they can't begin to understand what others are going through. I have also been told that I want a designer baby, which is as far from the truth as you can get."

Some mothers have remarked that once you have gone through treatment you will be done with worrying about what other people say to you. That is not to say that the comments still don't sting, but they are just not as painful. This is particularly true if the attempts at sex selection were successful. One mother says she was able to put the negative comments out of her mind when she thought of her baby girl. Before her daughter's birth, she felt it was more difficult, because the dream wasn't yet tangible.

Finding Support

Mothers who are thinking about sex selection have a hard time realizing that they might have a need for support. They want to talk to others, but fear saying anything because the threat of public "flaming" is still vivid in their minds. One mom doing preimplantation genetic diagnosis shares:

> "Part of it is that the court of public opinion has already voted, and mothers who want to do sex selection are guilty—but of what, I don't know."

It is important that you feel that you have the support for the journey and for the specific path that you've chosen. Making sure you have someone in your corner, or have the ability to talk to someone who has been there, can take your mind off many worries and free you to concentrate on the process. There are a variety of ways to find this support and myriad places to look for that support.

Find Support from Others Who Have Been There

Begin by asking at the fertility center if they conduct a support group of mothers who are going through the process or are considering it. If they don't have such a group, ask whether you can be paired with a graduate of the program, just for that moral support. You might also offer to start a support group yourself. This might be a good way for them to promote the mental health of couples undergoing treatment at their facility.

Turn to the Internet. Multiple groups thrive online. Each group has its own dynamic. Some are focused on one aspect of treatment, such as MicroSort®, the Ericsson Albumin Method, or the phase that they are in (like trying to conceive a baby girl, or enduring a two-week wait). These groups provide a wealth of support and knowledge from people you'll likely never encounter face-to-face.

Some of the forums online are public, which means that nearly anyone who stumbles on them can read them. To safeguard your privacy, choose an ID on the forum that doesn't give away your real identity, should you not want everyone at your work or in your religious congregation to know what you are doing. You can also pay to join other private groups.

Talking to Your OB/GYN or Midwife

Many women wonder, after choosing sex selection, whether or not to talk to their obstetricians or midwives about their choices. Some mothers choose not to tell even their care providers, citing the fact that there is no reason for them to know, that it will not change in any way how their prenatal care or birth is handled.

But this approach can leave you in a bind, says one mother who did preimplantation genetic diagnosis:

> "I didn't even know what to say to my OB when she suggested genetic screening because I'm over 35. Because I obviously know this baby is fine, but how do I say that to her without telling her what I did?"

This kind of response is not uncommon in women who have used sex selection. The baby that they worked so hard to make is also a secret they will fight tooth and nail to keep quiet. This extends to even trusted doctors and longtime midwives.

It could be advocated that the provider you choose ought to know. They do not *need* to know because there might be something wrong, but they simply *ought* to understand where you are coming from in your prenatal care. The lengths to which you have gone to conceive a baby of a specific sex might make them more sympathetic to your needs and concerns if it comes time to do an ultrasound in the middle of your pregnancy, a procedure that will likely answer your questions about the sex of the baby you are carrying.

Let's say you show up for that ultrasound, dying to know the results. Your baby decides that today is a shy day and sits with its legs crossed. You leave without the information you have been waiting to hear for 20 weeks. You are in tears. Now you feel that your doctor really thinks you are crazy, because she doesn't fully understand everything that you have emotionally invested in the process as well as in getting the answer.

How comfortable you feel with your practitioner may be your guide when deciding whether to talk to him or her about the sex selection you chose. JoJo relates:

> "If I were at my old OB/GYN's office, I would feel comfortable telling the doctor that I've seen for the past 15 years what I'm planning, and I'm sure she would give me some good advice. Since we have moved and I am seeing a different doctor, I wouldn't approach him with this info, since a lot of people think it's wrong."

Many of the decisions we make about talking to our care providers are based on trust built over the course of the relationship. It's important that you find a practitioner with whom you can have that trust. If you find that you are lacking that feeling of being able to trust your doctor with the simple fact that you used medical technology to have a girl or a boy, perhaps you should be looking elsewhere for a doctor or midwife that you can trust enough to talk with openly and honestly.

There will be doctors who do not understand. Some will even be against the use of sex selection. You can try to probe for your practitioner's views before having the conversation with them. Just begin by asking some of the simple questions. Consider asking them what they know about the topic. If they respond with an enthusiastic description of what methods are available, then you know you have found someone in whom it is emotionally safe to confide. If they give you a weird look and tell you that it's only science fiction, you have your answer there too.

The majority of doctors and practitioners will fall into some middle ground. They may have only heard what the average American has heard on television or read in popular magazines about sex selection. They may believe that we're much farther away than we really are from finding a solution to the dilemma. And they may be thrilled to be told what you've found out. Says JoJo:

> "Lots of doctors are not familiar with Shettles or supplements or anything to sway the odds, and usually will tell their patients not to waste their time.

She concludes, obviously frustrated:

> I even mentioned fertility monitors to my old doctor and how they can be used for sex selection and she just made a face. Lots of people are just not willing to 'go there.' It's like it's taboo."

This most likely also means that they are probably not up-to-date on the high-tech solutions to this problem, either.

11

Financial Concerns of Sex Selection

Can you afford the costs of doing accurate sex selection? This is a question that thousands of couples ask themselves every year. The more accurate question, though, is can you *not* afford accurate sex selection?

When you are dealing with sex selection and your dream of having a baby of a specific sex, how can you emotionally afford not to find a way financially? You can do some high-tech methods with better odds than natural (50/50) by swaying your odds to 70 percent or more, with as little as $600. And let's not forget that typically the more you pay, the better will be your chances of having the baby of your dreams. Even with a round of in vitro fertilization combined with preimplantation genetic diagnosis and MicroSort®, you can minimize multiple tries at having the baby of your dreams probably for under $20,000.

Diane points out one example:

> "The price tag on sex selection can get hefty, but it's still cheaper than having multiple children trying to get a desired sex. I had a neighbor that was one of ten children—nine girls and one boy. Her mother said she would have as many as it took to get a Floyd, Jr."

When you add up everything that goes into raising a baby today—the clothes, the food, the tuition, the health care, and everything else—it is truly very expensive. One estimate is that it costs parents $10,000 to raise one child in the first two years and

$8,000 for each subsequent child. This does not even include the cost of the birth and the rest of the child's life, nor does this figure include college tuition.

Ignoring the financial cost, denying yourself the ability to use sex selection carries a hefty emotional toll. Many mothers report great depression and sadness over not being able to experience the relationship they want with a child of a certain sex. Some women are trying to replace a great relationship that they had with their mother or other female figure in their lives. Other women are electing to have a baby girl as a way to repair the thoughts of a broken relationship in their past. Either way, some feel that other children may be well cared for but deprived of a fully functioning mother because of her obsession.

Certain fathers also have very strong thoughts of having a certain sexed child. A father who lacks the son or daughter that he wants may face ridicule in his social circle or may simply feel as if something is missing, as many mothers report. Men seem less willing to discuss their personal opinions on sex selection, saying that it is really up to their wives or partners to make that decision.

Why Sex Selection Is the Cost-Effective Option

While some may give sex selection a bad rap, others are out there who are interested in pursuing this form of family balancing for their families. In fact, in many ways, one can look at sex selection therapies as the most cost-effective method of achieving family balancing.

Sex selection can reduce the number of children in a family. Some couples simply will not stop having children until they finally have the child of the sex that they desire. It does not matter whether they weigh the pros and cons of a larger family versus sex selection, it's just a fact of their preference. This is something that must be taken into consideration.

The emotional and financial strains of having more children than you originally planned can be tremendous on the family. Although they may not be insurmountable, they can cause problems within the family. Parents who are obviously struggling to deal with the emotions of something they don't have, but want, can also cause a strain on the kids already in the family.

Society is partially to blame for the sex selection phenomenon. If you look at a woman, pregnant with her second baby, how does society expect her to feel about the next one? She's supposed to desire a baby of the sex she doesn't have.

Let's say a family reports, "We can really only afford to have two or three children, and we would really like at least one son and one daughter." If they wind up with a child of the same sex for both number one and number two, they must then decide whether to try again with random odds or risk denying themselves the son or daughter they are dreaming of. So the question becomes, do they throw caution to the wind and give nature one more try for number three?

If number three turns out to be another baby of the same sex as the other two, what then? They are either going to give up their dreams of having a baby of the opposite sex or they will financially risk having another baby. They could obviously keep on going, until they run out of money, time, or energy.

The other option is for them to invest the money into a highly effective sex selection technique that will either drastically increase their odds of having the baby of the desired sex or virtually guarantee it. The costs of raising children has gotten very expensive, and very few couples are in a financial situation to be able to afford large families. Should this alone be a reason to deny them a balanced family?

Technically you could have five children and never have a balanced family. But with the cost of each child estimated to be near a quarter of a million dollars apiece from birth till adulthood, the price tag of a large family can be steep, financially as well as emotionally.

Children need things that cost money, even when you are budgeting wisely. Sure, breastfeeding and cloth diapering can save you some money, but you still have to put clothes on the child's back and send them out into the world. A family that uses sex selection can reduce financial strain by opting out of having a large family while still having both boys and girls in the mix.

Sex selection can reduce the number of children with some genetic defects. A family that uses technology to accurately help it avoid having a child who will be born ill or become ill can also avoid physical and mental pain and suffering for all concerned. This reduces the financial strain on families, insurance companies, hospitals, and other agencies that provide physical and monetary support to these families. In short, it also saves society time, pain, and money.

We know that depression and divorce rates run higher in families that experience chronic illness, or the death of a child. One mother points out, "If we can help prevent some of this within a family by using sex selection, are we not obligated to do it?" Sex selection seems to be less an issue in such a circumstance.

Sex selection is less expensive than adoption. One mother pointed out that she was told to adopt a daughter. She was thrilled, because that thought had never occurred to her. This solution would seem to take care of multiple problems: her desire to mother a daughter and the adoption of a baby girl who was not wanted.

She began her search for a baby girl outside the borders of the United States. But when she and her husband sat down to add up the financial aspects of the adoption, they were amazed. "It was going to cost more to travel to adopt a baby girl than to do two rounds of MicroSort®," she says. In the end, she wound up not adopting and hopes to begin her MicroSort® cycle in the near future.

This cost analysis will not be the same for all families, for a variety of reasons. But the point is that perhaps sex selection will be less expensive than adoption. You should explore all avenues before making your decision on how to bring your child into your lives.

One mother was convinced that adoption was the best option to bring a daughter into their life. Then she started the process:

> "The scrutiny was awful. I felt like I was living under a microscope. I figured spreading my legs and having my husband's sperm under a microscope was less invasive, so we actually went back to do sex selection."

Ways to Pay for Sex Selection

Life is expensive. We all have bills to pay: the car, the house, groceries, school books, clothes, on and on. Our budgets contain categories for the things we must have, like electricity and water. But we usually have other sections that are not necessities, like cable television, movies, and recreation. Other areas might not be so cut and dried, like the cell phone or fancy haircuts or vacations.

While different methods of sex selection have varying costs from several hundred to several thousand dollars, only you can decide what your budget will allow. You will be the ones to determine how to go about paying for the treatments, which ones to try, and how much you can afford.

Some people want to do sex selection so badly, and feel it is so important to them, that they have written it into their long-term budget for several years. They slowly save up the money to afford the costs of having the baby that they are dreaming about, even if it means postponing pregnancy for a bit while they save the money. This kind of careful planning is not common in today's "I want it now" society.

Cash Payments

Paying with cash is certainly the easiest way to pay if you can afford it. It will not cost you extra in fees or add extra in terms of credit card interest. Most fertility centers, though, will not take actual cash payments.

The majority of fertility centers do not accept personal checks. You can have your bank make out a certified check or bring money orders to pay your bill. This ensures that you will pay your bill and precludes the clinic's charging more to cover bank fees associated with insufficient funds.

You will need to ask when your bill should be paid. It seems that the most common answer is by day 1 of the cycle in which you start treatment. Every fertility center will have slightly different rules concerning payment. You would hate to have a cycle cancelled or delayed because of a financial issue. So be sure to have the information you need beforehand so that everything will go smoothly from a financial perspective.

One mother told me that she asked up front about a cash discount when paying for her fertility treatments to do sex selection. "Insurance companies negotiate with doctors all the time. I figured it couldn't hurt. All they could do is say no." And that was exactly what they said at first.

In the end, however, she did wind up getting a cash discount on some of her services. They actually discounted her on an ultrasound. She saved about $200, a tidy little amount she did not have to pay. So go ahead and ask!

Credit Card Payments

Credit cards are probably the most common form of payment in the fertility clinics. They are not risky to the fertility center financially, and they are quick and easy. You should double-check that your fertility clinic accepts credit cards for payment. Some choose not to accept this form of payment because of the added fees they pay on the money collected in this manner.

You will have to figure out whether you have enough available credit to use your credit cards. You are allowed to break the payments up over several cards. "I used my credit card to pay for the medications," says Amanda. "Then my husband's credit card paid for the actual intrauterine insemination. That way he could say he got me pregnant!"

While you will have to pay the interest on the credit cards if you can't pay the total in full on your monthly bill, most couples find the convenience worth the price of the interest. Some couples told me that they opened up new lines of credit cards with low or no interest for a certain period of time, simply to do sex selection. They were able to pay off the balance within the required time without penalty and with no interest—like a short-term, no-interest loan. While this could be a great solution for many couples, be sure you read the fine print before opening any line of credit.

Other couples take longer to pay off the balances, but it allows them to start sooner with their treatment plans. Only you can decide if this is a good option for your family *and* your credit rating.

Taking out a Loan

Other couples choose to take out a loan to pay for their treatments and sex selection. This carries the same risk as other loans. To receive the most credible loans, you also need to have some form of collateral and a decent credit rating.

Loans come in many forms. Some are actually second mortgages on your home. These loans may be simple, one-time loans or lines of credit that can be used and reused as you pay into them and free up that credit again.

Other loans are straight-up. One dad laughs about his child being born through the use of a home improvement loan:

> "They laughed when I told them my definition of 'home improvement,' but our daughter has improved our home!"

Loans can be risky business, of course. Be very careful about overextending yourself. This also applies to offering collateral that you may not really want to offer. Some loan programs have horrible interest rates and stringent rules. Check with your Better Business Bureau for companies with negative reputations. Avoid what seems to be too good to be true.

Flexible Spending Accounts

The flexible spending accounts (FSAs) that many companies offer their employees might be just the ticket for you. This is money paid yearly into an account in your name. The money is placed there before taxes, so at a savings to you, to be used only for medical purposes. Flexible spending accounts are often used for copays, deductibles, or other noncovered expenses like orthodontics, glasses, and similar fees.

Contact your plan administrator and inquire whether your treatment plan and sex selection can be reimbursed through your flexible spending account. It may allow some things to be covered but not others. Since every plan is different, there's no way to tell in advance that you have a good chance of being covered. Using an FSA wisely might also be an easier way to save up for treatment, though it would require planning and coordination.

Insurance and Sex Selection

Most insurance companies do not cover sex selection in any form. The only aspects that may be included in your coverage would potentially be charges related to infertility. Some insurance companies will cover certain forms of testing, which you could use in the preliminary phases of preparing to do sex selection. An example of this would be blood work to test for communicable diseases like HIV or hepatitis.

To find out what your insurance company will cover, you should talk to them. When calling your insurer you should have your insurance card, your personal information, and the personal information of the person listed as the main member. This person may be you or your spouse. Typically they need the name, date of birth, social security number, and perhaps the place of employment.

You will want to ask them generically what your policy covers in terms of specific testing. This information will come from your practitioner, who can give you special codes to indicate a specific test and one for a diagnosis. Your insurance company may cover the test but only for certain diagnosis codes. So you might hear that your insurance company will cover an intrauterine insemination (IUI), but only for infertility-related diagnosis codes. This means that having it done simply for sex selection will not be covered. This is why both codes are important to have.

Depending on the staff abilities at the fertility clinic of your choice, they may be able to call your insurer for you. This can be a great option, because the staff will know what to ask and what to say. It is always possible for you to do it too.

Never ask your doctor to change your diagnosis code. This is insurance fraud. It can cost you and your family your medical insurance, create fines, and add other negative consequences. It can cost your physician the same penalties, but on a much greater scale. Insurance fraud is a huge drain on the economy. By tricking your insurance company into covering something that is not written into your contract, you not only run risks for you and your practitioner, but you also use resources that were not allocated to you.

If your insurance company does cover some of the expenses, it will be up to the fertility center of your choice to determine how payment will be made. Much of the fertility business operates on a cash basis. This means that some centers want payment up front.

Since insurance companies never pay up front for services, you would first pay the bill at the fertility center, then submit the claim to the insurance company. The insurer would then decide what it was going to pay and would send a check to you as reimbursement. This process depends on how quickly the fertility center staff files the claim and how quickly it is processed at the insurance carrier.

For couples who have infertility issues and have insurance coverage, what gets paid will vary widely. Some insurance companies spell out in very fine detail what they will cover, and for whom. For example, your insurance contract might say that it will cover in vitro fertilization for women under 40, but only for two cycles.

Some states have mandated infertility coverage, while other states have not. This leaves many families without much insurance coverage, even in an era of growing infertility disease. The fertility clinic will be helpful at working to get you the best coverage. Still, you will most likely have some out-of-pocket expenses. What they are can usually be determined to some degree before your cycle. This allows you to make the necessary financial arrangements.

Separate from fertility issues are genetic issues. Genetic testing may or may not be covered. Some insurance policies may require that you and your partner have genetic testing and counseling before allowing you to receive services for conception.

If you have been denied coverage from your insurance company that you believe you're entitled to, you always have the right to appeal its decision. An example might be that you're a carrier for a genetic disease that would be passed along to a baby boy and therefore want to use sex selection to have a baby girl. Your insurance company says it will not pay for sex selection techniques, including preimplantation genetic diagnosis.

So you could write the company a letter, including information such as why you want to avoid passing this disease on to a child. You would talk about not only the potential pain and suffering of the child, but also the medical costs associated with the disease. Explain that allowing monies now for sex selection to prevent this disease from occurring will save the insurance company money in the long run. This may be something that will cause them to at least reconsider. One mother says that she believes that the more you write and press your case, the more likely you are to win and at least get *some* coverage, if not all that you'd wish.

Even if the insurance company does decide to cover some or all of the procedures associated with the process, this does not relieve you of your burden to pay your originally agreed on portions of the treatments. You'll still be responsible for all copays, deductibles, and so forth. Your fertility clinic and insurance company can let you know how much that will be before you begin a cycle.

Retirement Funds for Medical Needs

Some retirement funds can be used for medical needs. Since infertility is considered to be as much a disease as is a sex-linked disorder, you may be able to withdraw retirement funds (taxable) to pay for expenses related to sex selection if you suffer from either of these problems. Other issues may be taken on a case-by-case basis. It certainly can't hurt to talk to your retirement fund administrator about all the issues surrounding the use of your retirement funds and the potential financial and tax liabilities of doing so.

Research Grants and Other Monies

Some programs have been set up that can actually provide a financial break for all or part of certain treatments. An example would be the MicroSort® trials that offer a discount on fees for the actual XSORT® done for couples who have X-linked disorders.

You may still need to pay money out of pocket for other expenses. This might mean you cover the tab for medications, or pay extra for the preimplantation genetic diagnosis, if you decide to go that route. Anything covered by a grant or a program, of course, is that much more you don't have to pay yourself.

Unfortunately, these programs are few and far between. And even when they do offer breaks, many families apply. Therefore, this should always be something on your radar to look out for when considering sex selection.

In the end, one MicroSort® mom says she recommends the following:

> "We did a home equity loan, as do lots of other mothers. I think a lot of women would do it if they had the money. Beg, borrow, or steal…"

Are There Societal Costs?

Some societies or countries have banned sex selection, claiming that the risks far outweigh the benefits. Yet the countries that currently have outlawed sex selection haven't managed to stem the flow of those who desire it. It is in countries like these that we still see the sad crime of infanticide.

As you may know from personal experience, some mothers and fathers go to great lengths to secure the funds to be able to choose the sex of their next baby. Some women work extra jobs in the middle of the night to save up funds for treatment. Or they work when their husbands are at home, to avoid paying for babysitters. This means they can accumulate the needed money faster and start receiving treatments sooner for the baby they so desire.

One family was selling household goods on eBay as a way to earn some extra money, all in the hopes of having some extra cash to be able to afford MicroSort®. This mom felt a pinch because of the age cut-off for this technology, and said she needed to hurry.

So there are many families around the country who are answering the question of whether or not sex selection is affordable. One desperate mother summed it up in this time-honored way: "Where there's a will, there's a way."

12

Finding out the Sex of Your Baby

"It's a boy!" an excited father cries out. While this sounds like the scene from any birthing room, it's really just the ultrasound room at a local midwife or obstetrician's office. By many accounts, the overwhelming majority of today's parents choose to learn the sex of their baby before birth. While parents do this for a variety of reasons, some of the more frequently cited ones are as follows:

- Buying the right clothes or painting a room the right colors
- Planning their house—how many bedrooms will they need?
- Preparing an older sibling for the arrival of the new baby
- Preparing themselves, as they don't like surprises
- Why not?

When you add sex selection, or even just sex preference, to the mix, that number is much higher. The majority of mothers (and most fathers) who have a specific desire for the sex of their baby will attempt to find out before its birth. While all the above reasons may also be true, the underlying theme is that parents just don't want to be disappointed when their baby is born. S. M., a mother of three boys, reports:

> "I have spent a lot on this process. The emotional energy. The physical energy. And certainly the money.... I need to know—like, yesterday."

Then there are couples who take a more low-key approach to finding out whether or not they were successful in conceiving the baby of a certain sex for which they hoping. Betty gushes:

"I found out with both of mine and was ecstatic each time. I never had a 'feeling' with our daughter other than joy that we were finally pregnant. When I was pregnant with my second child, I just dreamed and thought constantly it was another girl, but the ultrasound confirmed it was a boy. Deep down inside I wanted a boy but was afraid to hope for one because it seemed selfish."

She admits:

"We always wanted to know the sex if possible. We decided that if the baby cooperated during the ultrasound then we wanted to know; if not, it wasn't meant to be and we would have to wait. Both times we were happy—tears and all!—and excited about the new life we created and the journey that lay ahead for us. One of each!"

Finding out the sex during pregnancy may be less an issue with couples who have chosen to go the preimplantation genetic diagnosis route, where only eggs of the desired gender are implanted. For those who are using technologies like MicroSort® or low-tech methods of sex selection, there's always the chance that you will have a baby of the opposite sex from that which you planned. This makes determining the sex an important discussion to have with yourself, your partner or spouse, and your practitioner.

Questions to ask yourself before going ahead with any procedure for the purpose of sex determination include:

- Will the test or procedure give us the information we need?
- What is the accuracy rate of the information provided?
- Does the test or procedure have any risks to my health or the health of my baby? What are those risks? How likely are they to happen?
- When can the test be done? When are the results given to us parents? Will we find out that day, or will we be expected to wait? If so, for how long?
- How are the results given?

If you choose to find out the sex of your baby before birth, here are some tips to keep in mind when finding out.

Bring someone with you who understands your concerns. Do you want that person to be your husband or partner? Or maybe a friend who would better provide the support you need? No matter whom you choose, be sure that the person understands all the emotional energy you've put into the pregnancy and how the positive or negative outcome might affect you.

This person needs to be someone who can hold your hand whether the news is the news you were hoping for or other news. They need to be nonjudgmental. This can be a problem if you are taking your spouse and they don't happen to be of the nonjudgmental variety.

If you do not plan to take your spouse, how will you handle that? Does he not want to know? Or are you afraid he can't handle the whole thing? With our last baby, to give a personal example, my husband didn't wish to know the sex of the baby. I did. While I was in the waiting room we kept text messaging back and forth about the subject (so as not to disturb the other people waiting). He knew that I had a preference. He kept trying to interpret the way I was talking or what I was saying as news, but I refused to tell him. Actually, though, I got a kick out of his guesses, and why I thought he was making them.

Decide what your expectations are beforehand. What do you expect from the procedure? Do you want answers immediately? Be sure to share them with the person performing the procedure. They can help you adjust your expectations if needed—before you begin. Or they may be able to provide you with a better experience if you explained a bit about your background. This might also help the technician avoid saying generic "he's" and "she's." One dad explains:

"I just knew the baby was a boy, even though the tech said she couldn't tell. The tech used 'he' and 'him' throughout the ultrasound. I was squeezing my wife's hand. I knew she had to be so disappointed that everything hadn't worked out. You could have peeled me off the ceiling when she announced that the baby was a girl!"

Not every mother receives the news that she hopes for. Sharon, a mother of three sons, has this to say:

"Prior to getting married, I went to a psychic and she told me I would have two boys and that I was very lucky because boys are so much easier than girls. So, when my first two children were sons, I knew that this last one would have to be a girl. Otherwise, the psychic would have said I was going to have three boys, right?

So, when we got the first ultrasound, I saw it as plain as day—the penis I thought for sure would not, could not, be there."

A startled Sharon did what anyone else would have done:

"I started to cry. Not loudly, just silent tears streaming down my face, into my ears since I was lying down. Because that was when I was sure that the baby wasn't going to live. After all, the psychic was right about the other two boys...."

Sharon does admit that it wasn't the reaction she would have planned, "It seems so silly now, but you know how hormones do rage during pregnancy."

Decide how you want to be told. Do you want to find the sex first yourself, from the ultrasound image? Do you want to be told orally? Would you prefer to have the information written down and sealed for reading later, at home? Whatever method you choose, be sure that you let the technician know before you begin, and remind the technician as you proceed if you aren't finding out immediately.

One thing that some couples do is try to make the discovery a special, personal thing between the two of them. One mother tells of her plan to make this a moment just between her and husband, rather than between the two of them and the ultrasound tech.

"We had her write down the sex of the baby on a card. We brought an envelope that was thick enough that you couldn't read through it. The technician was

> asked not to use pronouns or to say anything that might alert us. We even went as far as having her make us look away during the sensitive periods of the scan.
>
> I didn't want to embarrass myself if I cried from disappointment or joy. But there was something special about opening the envelope together and reading the words. It just seemed more intimate—no matter what the answer was."

Know what you're getting into. Learn the risks, benefits, and alternatives of the procedure that you're choosing so that you can make the best decision for your family. Is it worth the risk of harming the pregnancy to find out ahead of time what the sex of your baby is? Some families say yes, that knowing the sex of their baby is worth the risk of some of the high-tech genetic testing available. R. S., who had been trying multiple low-tech ways to conceive her first son, says:

> "We knew that we wanted a healthy baby, though we obviously had a preference for a boy. We chose to do an early amniocentesis when our AFP came back slightly elevated. We tried the level II ultrasound and they said the baby looked good with no physical markers like a heart problem, but they couldn't determine if the baby was a boy or not. I don't think I would have chosen an amnio based on the AFP results alone. Their not being able to tell if the baby was a boy or not really pushed me over the edge. It was worth the risk to know conclusively."

Ask questions. Before the procedure, ask all your questions. Write them down ahead of time if you are worried that you won't remember them. This ensures that you will get them answered. You may also want to write down the answers, so you will have them to refer to later in calmer moments.

In the days or weeks before you go for whichever procedure you have decided on, play the results in your mind. This mental rehearsing will help you prepare for the news—either way. How will you celebrate if your sex selection efforts have been rewarded with the news you're hoping for? Will you cry with relief? Will you just lie there and silently take it in, not believing what the technician has said? One mother of two girls remembers:

"We had the tech write down the sex of the baby on a slip of paper. Since my husband didn't want to know the sex, we agreed we'd open it right after the birth, just to see if they had guessed correctly. So the tech hands me the envelope and I realize it isn't sealed. When I slipped into the bathroom, I had to look at the paper. She had stuffed a picture inside with the words 'It's a boy!' on it. I was so happy and relieved; I just leaned against the wall and cried. I was able to relax a bit and I even managed not to tell a soul about it. I also sealed the envelope before I left the bathroom."

Finding out the sex of your baby is a very special moment. Some couples don't want to hear it shouted out in a room, or even to share that moment with another person. You may choose to have the information written down. Then you and your partner can do with the information as you please. A. R., mother of two boys, recalls:

"I knew I would cry and be horribly sad if the news wasn't what I was hoping for. So we had the ultrasound technician write it down and we took it home to open it in private. It was a very nice and intimate moment for us. And neither of us had to worry about our reactions, as we were safe together. Fortunately it became a big celebration! My husband even surprised me with a bottle of sparkling cider. I will never forget that moment, and it would never have been the same if we had found out during the ultrasound."

R. S. recalls her last ultrasound:

"Our nerves were really frayed. We had put so much into this and I really needed it to go our way. My husband and I both just felt in our hearts that we were carrying the boy we had been hoping, praying, and trying for for so long. I made the decision to go to the ultrasound alone. When the technician asked me during the ultrasound how many girls I had at home, I told her six. She said, 'I thought this was your sixth baby.' I had been there before and knew the hamburger sign of three white lines. Without being told, I'd already mentally calculated this—my sixth daughter—and added her to the list. Later this memory would help me realize I was okay with it, but at the moment I was so disappointed. My heart felt so heavy. I was trying very hard not to cry in front of these strangers. We had been so sure—how could we have been so wrong? I felt betrayed by my body."

Accurate Methods of Sex Determination in Pregnancy

There are a handful of medically recognized ways to find out the sex of your baby before birth. Each method varies in the procedure: how soon it can be done, its accuracy, and the costs. Some of the methods also involve risk to the pregnancy.

Ultrasound

Many healthy pregnancies use an ultrasound examination to scan for fetal anomalies or screen for possible problems with the baby. The most common point at which to have an ultrasound routinely done is around the twentieth week of gestation. At that point technicians can usually tell your baby's sex with some degree of accuracy.

Trying to use it earlier in the pregnancy can result in inaccurate results. The eighteenth week is as early as I would recommend getting an ultrasound for determining the sex. If your insurance will pay for only one ultrasound, I would postpone it until the twentieth-week mark if possible.

If you wait until later in the pregnancy, you may have more difficulties trying to ascertain whether the baby is a boy or a girl, simply because there isn't enough space for the baby to stretch out and be completely visible. This is particularly true with multiples. Talk to your doctor or midwife about their beliefs and understandings. Tell them that your primary interest in the ultrasound is the determination of the sex of the baby and how important that is to you.

The accuracy of sex determination by ultrasound varies. Some ultrasound technicians and practitioners claim to have achieved 100 percent accuracy in their predictions. While this may be true, we have all heard stories of those women who went in for a test, had their baby, and found out at birth that the ultrasound had predicted the sex erroneously.

One mother shared with me that she had been told by numerous doctors and technicians that her baby was a boy. Because of a high-risk pregnancy, she had had numerous ultrasound scans. At the end of her labor she was amazed to hear the words, "It's a girl!" She says:

> "I thought about it for a minute. Then I looked at my husband and said, 'He's my son and I love him.' It was a weird realization after all those months of wanting and planning for a baby girl. But I think that in the end it was fine. Thinking I was carrying the girl I wanted helped me through a really rough pregnancy. Once he was born and we were both healthy, I realized how happy I was to have a healthy baby."

She later went on to have a girl using the Shettles Method.

Just how likely you are to get the correct answer will depend on many factors. One of them is your stage of pregnancy at the time of the ultrasound. As we've already discussed, the perfect window is between 18 and 22 weeks of pregnancy. Other factors can hamper or hinder your efforts at determining the gender of your baby.

Your baby itself will even play a part in determining the sex via ultrasound. If your baby is turned in a manner that makes it difficult to see the area that the technician must scan, you will have a hard time getting any reading, let alone an accurate one. An additional issue with your baby is that he or she may be "shy"—meaning they have got their feet tucked into their genital area and will not spread their legs for the scanner. Sometimes jiggling your belly or getting up and moving around will help get the baby to be a bit more cooperative, but then again maybe not.

Other maternal factors are also involved here. If you are overweight or obese, the machine may give a less clear picture. This is because the sound waves from the ultrasound machine have to travel through the extra-thick layers and return.

The skill and training of the technician or doctor who does the ultrasound will also matter when it comes to accuracy. Is he or she certified to do obstetrical ultrasounds? Or is it a nurse or medical assistant who has been doing them in the office without formal training? While it's not a guarantee, the longer they have been doing it, the better they will be at reading the more subtle differences.

The condition of the ultrasound machine itself can also play a part in how accurate your reading is likely to be. An older machine may be less able to give a good picture. The less detail, or the fuzzier the photo, the less likely it is that the machine will produce correct results.

If you decide that the ultrasound method of sex determination is right for you, you start the process by scheduling the ultrasound wherever your practitioner has ultrasound machines. This might be at their office, a hospital or birth center, or even an imaging center. Coordinate a day that is within the optimal window and a time when you can have with you the support people of your choosing.

To prepare for your ultrasound, you may be asked to arrive with a full bladder. Usually drinking about 32 ounces of fluid (four glasses) before the examination and not urinating will accomplish this. While this can be fairly uncomfortable for a pregnant woman, it's particularly effective at helping the ultrasound technician to visualize the contents of the uterus and your baby. You may be allowed to urinate after some initial measurements, or you may need to wait until after the entire procedure.

You will most commonly be asked to pull your pants down to hip level and lie back on an examination table. Your pants will be draped with a towel or cloth to prevent them from becoming coated in the transmission gel used to aid in the ultrasound. The technician will then use the ultrasound wand on your abdomen.

In some centers there is only one tiny screen that the tech uses for everyone to watch the exam. However, it is becoming more common to see ultrasounds done with the monitors hooked up to larger, additional screens aimed so that you don't have to crane your neck to see them. This allows you more comfort during the exam and also lets your support people watch more easily.

Typically the first things that the technician looks at is a standard order of items, such as head circumference, abdominal circumference, movement, heart rate, and so on. One of the very last things done is the revelation of the sex of your baby. That said, sometimes your baby is in such a position that the technician can see without much effort and will announce it earlier in the exam. Again, it is imperative that you talk to the tech before the exam begins. If you choose, it may be possible for her to tell you the sex earlier in the exam.

Another word to the wise: don't try to guess from what you are seeing on the screen. We've all heard the story of the couple who see an umbilical cord and think "boy!" Boy or girl—it doesn't matter—both can look very much alike in pregnancy ultrasounds. So let the technician show you screen shots of the genitalia and explain to you why she says what she does.

Typically you'll either see a penis and scrotum in the ultrasound or see what's described as three white lines, sometimes referred to as a "hamburger," representing the labia and clitoris of a girl. Sometimes a technician isn't able to see anything clearly and will make her best guess. Ask if this is what she's doing, because later you'll want to know. You'll most likely replay the entire examination in your mind over and over.

When the examination is finished, you can wipe off your abdomen. They usually have a bathroom nearby where you can relieve your full bladder. You may be given some photos or a video of the examination, or both, depending on the rules of the center. Then you're free to go home. It typically takes at least a couple of days for the doctor or practitioner to review the exam and provide you with any results, beyond the baby's sex. If there's a concern or a need for further study, your medical professional will usually let you know about it on the day of the exam.

The risks of ultrasound are extremely low. The majority of studies say that ultrasound has been shown to be safe, but nevertheless should only be used when medically necessary rather than frivolously. Your insurance company may have the last word on whether your ultrasound is deemed to be a medical necessity.

If your insurance company doesn't cover the cost of an ultrasound for the reason of sex determination or even for the fetal anomaly scan, consider whether you want to pay that cost out of pocket. The majority of practitioners' offices are aware of whether your insurance company will pay for the exam. If yours doesn't know, you can always call your insurer and ask yourself.

Should they not pay, you'll need to decide if this is something you want to cover yourself. If you go that route, don't be tempted to have a bargain ultrasound. While convenient places are cropping up in malls and in stand-alone centers that are set up solely for doing elective ultrasounds, the Food and Drug Administration (FDA) has warned against their use. Instead, opt to pay your doctor or midwife for the use of the time, equipment, and expertise. It may save you some headache and perhaps some heartache in the end. Typically you are looking at paying between $250 and $600 for an ultrasound, though the price may vary from practitioner to practitioner and even regionally.

The inaccuracy and limitations of the ultrasound still make this a somewhat chancy method of sex determination. This bothers some parents enough that they turn to other high-tech methods of sex determination.

Chorionic Villus Sampling

One high-tech way to determine sex with virtually 100 percent accuracy is called chorionic villus sampling (CVS). Its main goal, however, has always been to screen for genetic problems. Typically, chorionic villus sampling is offered to the following:

- Women over 35 years of age
- Women who have previously had a baby with one or more birth defects
- Couples with a specific family history of birth defects

This test can be done earlier in pregnancy than any other test currently available to determine the sex of your baby. It is usually offered toward the end of the first trimester of pregnancy, usually after the tenth week following your last menstrual period (LMP). The results of the test are not immediately available. It can take about 10 days for the results to come back, though occasionally they are available sooner. The procedure carries a small risk of miscarriage and is contraindicated if you have experienced any bleeding or spotting in your current pregnancy.

The procedure is done by having you report with a full bladder for an ultrasound examination of the pregnancy. You will lie on a table, usually in a gown or without your pants and underwear. Ultrasounds may be done with a transabdominal wand, as done later in pregnancy, or may be done with a transvaginal wand, which actually is inserted inside the vagina. This provides your doctor a better view of the pregnancy and the area where the sample will be taken. You'll have the area cleansed to avoid infection.

This sample may be taken through either a small tube inserted through the cervix or a needle inserted through the abdomen. The extraction method to be used depends on the anatomy of your body, the location of your placenta, and other factors. Typically the transcervical route is the most common and considered to have the lowest risk factors, unless you have a retroverted uterus, meaning that your uterus is tilted. If you do have a retroverted uterus, the abdominal route is usually the safest option.

The procedure involves removing a piece of the chorionic villi. This is tissue that adheres between the uterus and the amniotic sac. It is genetically identical to your baby. So when the doctor takes the sample, they can easily look at the genetics of the baby.

After the chorionic villus sampling, you will be cleaned up from the gel and the cleansing solutions. You will stay at the center for about an hour to monitor you and the baby. Your baby's heartbeat will be monitored occasionally during this time to ensure that your baby tolerated the procedure well. Once you are released to go home, you will be asked to take it easy for a few hours to a few days, depending on how you're feeling. You will be advised not to have sexual intercourse or to use anything in your vagina for a certain period.

The rate of miscarriage after a chorionic villus sampling is between 1 in 200 and 1 in 100, though recent studies show this might be even less likely. About a third of all women will experience some spotting for a couple of days after the procedure. About one in five women may also experience cramping afterward. Both symptoms will likely be self-limiting, though you should report them to your doctor.

Pregnancy loss rates and complication rates are directly tied to the experience of the doctor doing them. Be sure to find the person who has a great deal of experience in doing chorionic villus sampling. Ask about their complication rates and the number of procedures they perform on average, and then make a decision as to whom you will see.

The cost of chorionic villus sampling can range from $800 up to several thousand dollars. This would normally be an out-of-pocket expense, particularly if no medical reasons exist for conducting this test, such as a family medical history that indicates the need for genetic testing, or advanced maternal age, or even issues in the current pregnancy (like abnormal results from screening tests). Your health insurance may cover all or part of the costs of this test if there is a legitimate need for it.

Mary chose chorionic villus sampling for her genetic testing, after her successful MicroSort® attempt. She recalls:

> "Once I got pregnant, there was a fear of God punishing me for messing with nature. Maybe there is a reason I didn't have a girl. Is she going to have three eyeballs because I wasn't meant to have her? Once I knew it was a good outcome, then I told the world. My friends were happy for me."

Amniocentesis

Another option for both sex determination and genetic testing is amniocentesis, which may be offered for the same reasons as a chorionic villus sampling. The primary determiner of which test is offered to you is the timing. Amniocentesis is done later in pregnancy, typically between the fifteenth and eighteenth weeks after your last period, though some practitioners offer as early as the eleventh week of pregnancy. First-trimester amniocentesis is currently thought to be a bit riskier for the baby and the pregnancy than having one in the second trimester, which is the commonly accepted time frame.

You may routinely be offered an amniocentesis for your pregnancy if:

- You are over 35
- You have had a previous baby with a birth defect
- You have a known genetic disorder or a family history of one
- Prenatal screening (AFP, and so on) has shown you to be at a higher risk

With a full bladder you will first undergo an ultrasound examination. The technician is looking at your baby's general health, the position of the baby, pockets of amniotic fluid, and the location of the placenta. This is usually done with the transabdominal ultrasound, depending on the location of your placenta. Once a location is determined, your belly will be washed and marked for the placement.

Amniocentesis is performed with the continuous use of ultrasound guidance. This helps the practitioner place the needle through your abdomen into the selected pocket of fluid. Usually there is a second technician working the ultrasound. Together they watch how the baby is responding and moving, to avoid poking it with the needle.

The goal is to remove only a small sample of amniotic fluid, which contains skin particles form your baby. These particles are then used to extract the DNA to be able to give you an accurate view of your baby's genetic makeup, including its sex.

The needle is not pleasant to look at, so I would advise against watching the procedure if needles make you uncomfortable. The good news is that this needle feels no different than any other injection. Some practices choose to numb the skin first, though women vary on whether it helps. The numbing agents do burn. Talk to your doctor and other women who have had an amniocentesis to help you choose which way is best for you.

After the sample is taken, you will lie on the examination table for a bit longer. They'll monitor the baby's heart rate and then release you to go home. You should plan on taking it easy for a couple of days after the amniocentesis.

The risks with amniocentesis are real. For a second trimester amniocentesis, from 1 in 400 to 1 in 200 women will suffer a miscarriage from the procedure. These numbers are higher for women in the first trimester, which is why chorionic villus sampling is recommended when you do first-trimester genetic testing. Some 1 to 2 percent of women will have other symptoms, like spotting, cramping, or even amniotic fluid loss after the procedure. These should not last long and should be reported to your doctor or midwife. The risk of uterine infection from the procedure is small, about 1 in 1,000.

Undergoing either chorionic villus sampling or amniocentesis solely for the purpose of sex determination isn't recommended, because of the added risks to the pregnancy and potentially to the mother. However, if you will be having genetic testing done for a separate reason, both these methods provide nearly guaranteed accurate results of sex determination.

If you have medical reasons for having the genetics testing done, your health insurance may cover all or part of this test. You should talk about this with your care provider beforehand and check with either your benefits coordinator at your place of employment or your insurance company for the final say.

Other Methods of Sex Determination

There are many other ways during pregnancy that supposedly determine whether your baby is a boy or a girl. Some of these are considered to be old wives' tales, like the ring test. This is where you take your wedding ring, tie it to a string, and hold it over your pregnant abdomen. The ring will either swing back and forth, indicating that the baby is a girl, or it will go in circles, indicating that your baby is a boy. This type of sex determination is fun and silly. It's also quite obviously not to be taken seriously.

If you look at other methods, like combining urine and Drano® for sex determination, you might find yourself thinking that some might actually have a foundation in science. The truth is that this is not currently a way to find out if your baby is a boy or a girl. In fact, this method can be dangerous. A number of accidental burns and inhalation of fumes from the Drano® have been reported. Several studies have been done trying to prove this one to be accurate, but alas, it is not. A number of urban legends websites offer this information as well.

The real problem comes when you have very scientific-looking methods claiming to be able to tell you the sex of your baby quickly and accurately. One of these was the Baby Gender Mentor. You take a small sample of maternal blood as early as five weeks into the gestation, send the sample to the company's lab, and learn within a few days if you are expecting a boy or a girl.

This test sounds too good to be true. And indeed it appears *not* to be true. Several families have experienced this first hand. The pain and sorrow caused by the mistaken testing has not been pleasant.

For a price tag of about $300, it's also more expensive than your average old wives' tale or even a low-end ultrasound examination. With the lack of science and a sound practitioner behind it, the chances that it's good would be very slim. This business was threatened with legal action after many women and their families were told very negative and untrue statements.

What If You Don't Want to Know before Birth?

Not finding out the sex of your baby before birth is also an option. Yes, even when sex selection is involved with the conception, some families choose to wait until birth to find out whether their efforts succeeded. This is something that many families can't comprehend, but it is nevertheless a valid position.

So why would you do everything within your power to influence the sex of your baby— and then not bother to find out ahead of the actual birth? When I asked women that very question, the reasons that I got were varied.

One mother told me that she had put everything she had mentally, emotionally, and financially into her MicroSort® cycle. She already loved her baby, and while she hoped it was a boy, she would love a daughter as well. She said that the chances were in her favor and that she needed this thought to get through the end of her pregnancy and through labor. She'd deal with the outcome, either way. She says:

> "I need the thrill of labor. If at birth I find out that my baby is a girl, it will be only after I love her completely as my baby in my arms."

This mother echoed what others have said to me, that the hormones from the joy and work of labor would help even if the outcome was not what had originally been planned. Some of these families feel as if they need the lift that birth provides new mothers, to help deal with what might otherwise be something sad or disappointing for them.

One father explained that they had been very excited in a previous pregnancy when the ultrasound technician had told them that they were expecting a baby boy at about 20 weeks' gestation. That joy was turned into much physical activity in preparing a home for their eagerly anticipated son. During her cesarean birth, his wife said she felt disconnected in general, but even more so when the doctor held up a baby girl. "We were both just shocked…like, 'That's not our baby. We're having a *boy*.'"

After having that experience in their first pregnancy, they weren't willing to be burned again.

> "This time we've stacked the deck to have a son. Either way we've won with a baby, but hopefully we've completed our family with a boy," the father adds. "I'd hate to experience that elation and disappointment we experienced last time. It was so awful having to tell everyone our son was now a daughter."

When dealing with the disappointment of having a baby of the opposite sex, it is important to acknowledge this loss. This feeling has been referred to by some as a mental miscarriage. You are losing the dream of a baby of one sex or the other.

Some families don't opt to find out, simply because they can't. They may not have the insurance coverage or the money to afford testing, even ultrasound testing. This can lead to frustration and hard feelings:

> "I really felt like I needed to know, despite everything we did, and yet I'm stuck because of insurance," explains one mother. "I am going crazy here!"

Others cite the risks of some of the more accurate but invasive measures, like amniocentesis or chorionic villus sampling. They say that the rate of pregnancy loss, while small, is real. That risk is not acceptable to everyone.

Confidence is another factor. One mother asked me rhetorically why she should risk the time, the money, or the pregnancy just to see if her baby was the daughter of her dreams. "We chose her. I know it's my daughter. I just know it," beams A. E., a mother-to-be. "I don't need any test to tell me what I know. That's why we did the genetic diagnosis before putting this particular embryo in my uterus. We picked her specifically. How many parents can say that?"

No matter what choice you make in finding out the sex of your baby before birth, let's hope that it will be the right choice for your family. It should be a decision that you can easily live with—one that helps you prepare mentally and emotionally to add a new life into your family.

Living with Sex Disappointment

The truth is that it can be completely normal to suffer from disappointment at having a baby of the opposite sex for which you were planning. This is true even if you were merely conceiving in the old-fashioned way. But if you want to up the stakes by investing both the emotional energy and the financial outlay, then you may really be in for it when you find out.

Everyone reacts differently. The healthy baby card is one that people often talk about. Says one mother:

> "I'm glad the baby is healthy. It's not about that. How can I not be happy the baby is healthy? But can't I simply be sad that I am not ever going to have the baby girl I've dreamed of all of my life? What is so wrong with that?"

Her words ring true for many couples, particularly the women. The problem is that when these women are told that they should be happy that their baby is healthy, their true feelings are being ignored, which is both harmful and painful.

This can lead to myriad problems. These can include an increase in postpartum depression, particularly from repressed feelings. After all, if you keep telling a woman that her opinions are not valid, she will stop talking to you.

Some people report feeling selfish. "Having a child is something that not everyone can do. I created and gave birth to an amazing creature—and here I am selfish and wanting more. I should be ashamed of myself," exclaims the mother of a son, after a Shettles baby girl attempt.

You hear comments like the following:

"I mean, it's not like I don't have a lovely baby. What is wrong with me that I can't be happy with the baby I have?"

"I've always seen myself as the mother of a girl. I don't know what to do with a baby boy. It just wasn't a part of my plan."

Parents need the following things:

- Validation from others
- Time to heal
- Knowledge that they are not alone
- Screening for postpartum depression (fathers as well as mothers)

Trying Again

If a couple has failed in its first attempt at sex selection, the question may arise whether they should try again. This is a question that only the couple can answer. They need to make sure that they have the physical, mental, emotional, and monetary strengths needed to try again.

If the failure was the lack of pregnancy or a miscarriage, the answer may seem a bit more clear-cut: try again. Since it takes the average couple more than one try to get pregnant, you may have already decided that you would try a fixed number of times before giving up hope. Most couples have the drive and determination to continue.

If the "failure" was the birth of a child of an opposite sex, the question may not be as easy to answer. You have another baby in your life—can you afford another, in all the ways an additional child will need you? Some mothers say that they just couldn't stop, that they had to keep going. Talking to the practitioner at your preferred fertility clinic may be helpful.

While some couples do choose to go a different route, others are quite happy with the care they received and figure that their failures at the sex selection cycle were merely an odds game.

Choosing to do sex selection is never an easy road with no curves or dips. There are many decisions to be made and beliefs to discuss. You need to look at all your options and ask your family hard questions about how this new technology would change your lives.

You may decide that high-tech sex selection is the best choice to help you add that special baby to your life. If this is the route that you decide to go, you have multiple options available. Only you and your partner can decide which is right for your family, given your history.

So develop a plan, and then find the emotional and financial resources to put that plan into action. Have some back-up plans in case things go a bit haywire or you're thrown off course. Use the support system you have developed to help you stay strong. And finally, enjoy the precious new baby who has come into your life—the one you have specifically chosen.

Resources

Websites

Sex Selection Information and Support

In-Gender

www.in-gender.com

Girl or Boy? Sex Selection Techniques for Everyone before Pregnancy

pregnancy.about.com/od/boyorgirl/p/girlorboy.htm

Sex Prediction (including Old Wives' Tales)

Girl or Boy Quiz (Old Wives' Tales)

www.childbirth.org/articles/boyorgirl.html

pregnancy.about.com/od/boyorgirl/a/quizboygirl.htm

Girl or Boy Ultrasound Quiz

pregnancy.about.com/od/boyorgirl/a/boygirlquiz.htm

China Gold Lunar Calendar

www.chinagold.com/baby.htm

Sex Detection Via Ultrasound

pregnancy.about.com/od/boyorgirl

Pregnancy Websites

About.com Guide to Pregnancy and Birth

pregnancy.about.com

Childbirth Connection

www.childbirthconnection.org

Charting Fertility Cycles

Taking Charge of Your Fertility

www.ovusoft.com

Fertility Friend

www.fertilityfriend.com

Infertility Websites

About.com Guide to Infertility

infertility.about.com

The InterNational Council on Infertility Information Dissemination (INCIID)

www.inciid.org

IVF.com

www.ivf.com

RESOLVE: The National Fertility Association

www.resolve.org

Genetics Websites

March of Dimes

www.marchofdimes.org

Organizations & Fertility Clinics

MicroSort® (Genetics & IVF Institute)

www.microsort.net

Gametrics Limited

www.childselect.com

The Fertility Institutes, the Gender Selection Program
www.sexselection.net

American College of Nurse Midwives (ACNM)
www.mymidwife.org

American College of Obstetricians and Gynecologists (ACOG)
www.acog.org

American Society for Reproductive Medicine (ASRM)
www.asrm.org/Patients/topics/genderselect.html

Preimplantation Genetic Diagnosis International Society (PGDIS)
www.pgdis.org

President's Council on Bioethics Working Paper
www.bioethics.gov/background/background2.html

Society for Assisted Reproductive Technology (SART)
www.sart.org

Books

Sex Selection

Baby Girl or Baby Boy: Choose the Sex of Your Child by Mark Moore, M.D. and Lisa Moore (Washington Publishers, 2004).

Boy or Girl? The Definitive Work on Sex Selection by Dr. Elizabeth Whelan (Pocket Books, 1991).

Chasing the Gender Dream: The Complete Guide to Conceiving Pink or Blue with the Latest Sex Selection Technology and Tips from Someone Who Has Been There by Jennifer Merrill Thompson (Aventine Press, 2004).

How to Choose the Sex of Your Baby: The Method Best Supported by Scientific Evidence by Dr. Landrum Shettles and David Rorvik (Doubleday, 1996).

How to Choose the Sex of Your Baby: The Method Best Supported by Scientific Evidence by Dr. Landrum Shettles and David Rorvik (Doubleday, 1996).

The Preconception Gender Diet by Sally Langendoen and William Proctor (M. Evans & Co., 1982).

Getting Pregnant

The Everything Getting Pregnant Book: Professional, Reassuring Advice to Help You Conceive by Robin Elise Weiss, LCCE (Adams Media, 2004).

Taking Charge of Your Fertility: The Definitive Guide to Natural Birth Control, Pregnancy Achievement and Reproductive Health by Toni Weschler (Quill, 2001).

Infertility

How to Get Pregnant with the New Technology by Sherman Silber (Warner Books, 1998).

Pregnancy

About.com Guide to Having a Baby: Important Information, Advice, and Support for Your Pregnancy by Robin Elise Weiss, LCCE (Adams Media, 2006).

The Mother of All Pregnancy Books: The Ultimate Guide to Conception, Birth, and Everything In Between by Ann Douglas (Wiley, 2002).

The Official Lamaze Guide: Giving Birth with Confidence by Judith Lothian and Charlotte DeVries (Meadowbrook, 2005).

Acronyms, Abbreviations, and Definitions

The world of sex selection has a variety of terms and jargon. The following is a sampling of the most commonly used terminology both in this book and in other references. You may notice that some of these terms do not appear in the book: they are included here because you will frequently run into them, typically in an online environment.

2WW Abbreviation for "two-week wait," the time period after you ovulated or conceived until a pregnancy test can be taken.

AAP American Academy of Pediatrics

ACNM American College of Nurse Midwives

ACOG American College of Obstetricians and Gynecologists

AFP Alpha-fetoprotein test. This prenatal, maternal blood test is typically done between 15 and 20 weeks to measure alpha-fetoprotein levels; abnormal amounts may indicate a genetic disorder.

AI Artificial insemination. *See* IUI

AID Artificial insemination donor

AIH Artificial insemination husband

ANA Antinuclear antibody

APA Antiphospholipid antibody

ASRM American Society for Reproductive Medicine

Assisted Reproductive Technology (ART) A set of high-tech reproductive technologies, such as in vitro fertilization that help couples achieve a pregnancy.

Basal body temperature (BBT) Your body's baseline temperature, used to help detect slight shifts in your temperature without the "contamination" of movement or food.

ßhCG Beta human chorionic gonadotropin

CCCT Clomiphene citrate challenge test

Cervix (Cx) The mouth of the uterus that opens to allow a baby to be born.

Chorionic villus sampling (CVS) Early pregnancy genetic testing done to rule out genetic abnormalities. It can also tell you the sex of your baby.

Cycle day (CD) A day in your menstrual cycle, but used to count from a specific day. For example, you would call the start of your cycle Cycle Day One, though you might not ovulate until Cycle Day Fifteen.

D&C (dilation and curettage) A surgical procedure to clear the uterus of its contents.

DD Online text abbreviation for "dear daughter."

DE Donor eggs—eggs that are removed from another woman to donate to a different woman who cannot use her own eggs.

DES Diethylstilbesterol. A medication given to prevent miscarriage in the 1960s and 1970s that has caused numerous fertility issues for children born after their mothers took this medication.

DH Online text abbreviation for "dear husband."

DI Donor insemination—an insemination done using sperm from a donor for various reasons.

DNA Deoxyribonucleic acid

DS Online text abbreviation for "dear son."

E2 Estradiol

EDD Estimated due date. This date is calculated frequently from the first cycle day of your last normal menstrual cycle by adding 280 days. It is more accurate to use the 266 days from conception method. Due dates are estimates of when your baby will be born; most babies are born two weeks after or two weeks before this date.

Embryo transfer (ET) Placing the embryos back inside the uterus.

Endometriosis (ENDO) A condition in which the endometrial tissue grows outside of the uterus; this can potentially cause pain and/or infertility.

EPT Early pregnancy test. This test can be performed just slightly before your expected period.

FISH (fluorescence in situ hybridization) A type of DNA analysis used by MicroSort® in sex selection; the analysis indicates the percentages of the different sperm cells in a sorted sample.

Frozen embryo transfer (FET) Placing embryos back inside the uterus after having previously been frozen for storage.

FSH Follicle-stimulating hormone

Gamete intrafallopian transfer (GIFT) An assisted reproductive technology that takes the eggs from the woman's ovary and combines them with sperm from the male. The fertilized eggs are then immediately placed back into the fallopian tube. There is no embryo transfer step in this process.

Gestational surrogate (GS) A woman who physically carries a pregnancy for another who is incapable of being pregnant.

GnRH Gonadotropin releasing hormone

hCG Human chorionic gonadotropin, the hormone pregnancy tests measure to determine pregnancy.

HMG Human menopausal gonadotropin

Home pregnancy test (HPT) A pregnancy test done at home using urine to screen for the hormone hCG.

Hysterosalpingogram (HSG) A test in which dye is injected into the uterus during an X-ray to check for abnormalities of the uterus or tubal patency.

ICI Intracervical insemination

In vitro fertilization (IVF) A reproductive technology in which the sperm and the egg are retrieved and joined outside of the body before being returned to the uterus in the hopes of conceiving.

Intracytoplasmic sperm injection (ICSI) A procedure in which a single sperm is injected into a single egg. This is a treatment for certain forms of male infertility.

Intrauterine insemination (IUI) A procedure in which specially prepared sperm is placed inside the uterus at a certain time based on ovulation. This can be done to help increase pregnancy rates in general or because the sperm has been processed for sex selection purposes. Also known as artificial insemination.

IU International units

IVIg Intravenous immunoglobulin

Laparoscopy/Laparotomy (LAP) A surgical procedure done to correct fertility problems or to diagnose fertility problems.

LH Luteinizing hormone

LMP Last menstrual period

Luteal phase defect (LPD) A difference in your menstrual cycle in which the luteal phase, or second phase, is too short, making pregnancy difficult.

MESA Microsurgical epididymal sperm aspiration

MicroSort® (MS) A high-tech method of sex selection where sperm is sorted by sex using DNA to obtain sperm samples that are enriched by the sex you choose. This method can use IUI or IVF to help achieve pregnancy.

O Orgasm or ovulation, depending on context.

OB Obstetrician

OB/GYN Obstetrician/Gynecologist

OHSS Ovarian hyperstimulation syndrome, an unlikely side effect of using some fertility medications.

Ovulation prediction kits (OPK) A high-tech method to detect ovulation. These kits use your urine to screen for the luteinizing hormone (LH) that is released at the time of ovulation.

PCOS Polycystic ovarian syndrome

PCR Polymerase chain reaction

Pelvis inflammatory disease (PID) A disease that causes swelling and scarring along the fallopian tubes, which can render the woman infertile.

PESA Percutaneous epididymal sperm aspiration

Premature ovarian failure (POF) A condition in which the ovaries of a woman fail to thrive and begin their decline long before the average.

RE Reproductive endocrinologist

RPL Recurrent pregnancy loss

SART Society for Assisted Reproductive Technology

Semen analysis (SA) Screening for the sperm to test factors surrounding fertility.

STD Sexually transmitted disease. *See* STI

STI Sexually transmitted infection. A class of contagious diseases frequently spread through sexual contact. Also known as STD.

Testicular sperm aspiration (TESA) A procedure using a fine-needle biopsy to retrieve sperm.

Testicular sperm extraction (TESE) An open testicular biopsy to remove sperm.

TSH Thyroid stimulating hormone

Tubal embryo transfer (TET) *See* Zygote intrafallopian transfer

Ultrasound (US or U/S) An exam that uses sound waves to look at internal body structures such as the ovaries or even a baby.

UTI Urinary tract infection

XSORT® MicroSort® terminology for the X-bearing sperm

XX Abbreviation for a girl; represents the two X chromosomes needed to make a female gamete.

XY abbreviation for a boy; represents the X chromosome and the Y chromosome needed to make a male gamete.

YSORT® MicroSort® terminology for the Y-bearing sperm

Zygote intrafallopian transfer (ZIFT) A method of embryo transfer. Eggs are fertilized in a laboratory then inserted into the fallopian tubes. Also known as tubal embryo transfer (TET).

Index

About the Author

Robin Elise Weiss, BA, ICCE-CPE, CLC, CD(DONA), LCCE, FACCE, is a childbirth and postpartum educator, certified doula, doula trainer, lactation counselor, and proud mother of seven children (two sons and five daughters). She has two passions: pregnancy and writing. A natural fusion of the two has blossomed into a career of sharing her knowledge with others, using a blend of wit and wisdom in the classroom, through her books, and on the web. Since 1989, Robin has attended hundreds of births and educated couples about pregnancy and birth. She is the author of numerous books on pregnancy, including *About.com Guide to Having a Baby, The Everything Mother's First Year, The Everything Pregnancy Fitness Book,* and *The Everything Getting Pregnant Books.* Robin is also a freelance writer and lectures across the country on pregnancy and health-care topics. She hopes that one day she won't spend so much time driving the kids around in her minivan.

Acknowledgments

A special thanks to the mothers and fathers who have made the difficult decisions surrounding sex selection and willingly shared their stories.

Many people helped me with the birth of this book. Some of them are Marci Yesowitch Hopkins, Kim Goldman, Jana and Aaron Pedowitz, Paula Pepperstone, Diane Graf, Pat Predmore, Dr. Michael Grossman, Dr. Alberto Carrillo, Trina Leonard, and Barb Doyen.